Electric Circuits

Gengsheng Lawrence Zeng · Megan Zeng

Electric Circuits

A Concise, Conceptual Tutorial

 Springer

Gengsheng Lawrence Zeng
Utah Valley University
Orem, UT, USA

Megan Zeng
University of California, Berkeley
Berkeley, CA, USA

ISBN 978-3-030-60517-9   ISBN 978-3-030-60515-5   (eBook)
https://doi.org/10.1007/978-3-030-60515-5

This Springer imprint is published by the registered company Springer Nature Switzerland AG
The registered company address is: Gewerbestrasse 11, 6330 Cham, Switzerland

Preface

Are you a student who is looking to supplement what you are learning in class? Or are you simply interested in electric circuits? *Electric Circuits: A Concise Conceptual Tutorial* gives you an opportunity to understand fundamental electrical engineering concepts. This book is written in a reader-friendly format like a pictorial dictionary, and you can directly jump to any topic you want to learn more about without having to read the entire book sequentially. We hope that this book will help save your time in grasping difficult concepts in electric circuits.

Good luck and have fun!

Orem, UT   Gengsheng Lawrence Zeng
Berkeley, CA   Megan Zeng
2020

Contents

Voltage, Current, and Resistance

1

Electric circuits, such as the one shown in Fig. 1.1, consist of multiple connected electrical components so that electrons can flow through a closed loop.

Fig. 1.1 An example of an electric circuit, which is represented using a **circuit diagram**

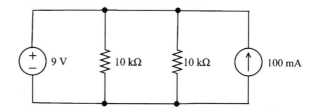

In order to analyze and design electric circuits, we must first understand some fundamental electrical quantities: voltage, current, and resistance.

Voltage is the difference in electric potential between two points in a circuit and is measured in volts (V). A typical reference point is **ground** (GND), which is a point we choose to be 0 V. However, it is also common to measure voltage across a component, as can be seen in Fig. 1.2. When expressing voltage as a variable, we usually use v.

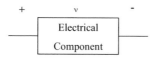

Fig. 1.2 Voltage across an electrical component. The "+" and "−" labels correspond to the positive and negative terminals of the component

Voltage across an electrical component is measured with respect to the negative terminal of the component instead of ground. This is equivalent to the voltage of the

G. L. Zeng, M. Zeng, *Electric Circuits*,
https://doi.org/10.1007/978-3-030-60515-5_1

positive terminal minus the voltage of the negative terminal, both with respect to ground.

To gain an intuitive understanding of voltage, let us imagine that you are hiking up and down a hill as shown in Fig. 1.3. We can consider the bottom of the hill to have an altitude of 0 and measure the altitude with respect to the bottom of the hill. As you go uphill, your altitude increases. As you go downhill, your altitude decreases. In this scenario, the bottom of the hill is like ground while the voltage is like altitude. Both the bottom of the hill and ground are reference points while both altitude and voltage represent differences with respect to a reference point.

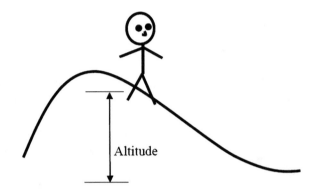

Fig. 1.3 A person hikes up and down the hill, which corresponds to changes in altitude

Let us say that there is a meadow on the side of the hill like in Fig. 1.4. If we want to determine the change in altitude of the meadow itself, we can measure the altitude of the meadow with respect to the bottom of the meadow instead of the bottom of the hill. This is akin to measuring the altitude at the top of the meadow, then subtracting the altitude at the bottom of the meadow. If we consider the meadow to be like an electrical component, the change in altitude of the meadow is like the voltage across the component.

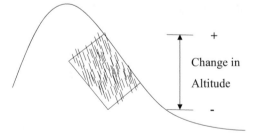

Fig. 1.4 The change in altitude of a meadow

Current is the flow of electrons through a circuit and is measured in amps (A). We typically use i for current as a variable. In a closed loop, voltage causes current. If there is no closed loop, current will not flow. Current through an electrical component refers to the current flowing through that component.

We can view the relationship between voltage and current as a closed-loop water system laid out on the hillside as shown in Fig. 1.5. There are two water tanks in Fig. 1.5: one at the higher altitude, corresponding to higher voltage, and the other at the lower altitude, corresponding to lower voltage. The water will naturally flow from the upper tank to the lower tank through the water pipe. The altitude difference h creates a gravitational force to push the water to flow from the upper tank to the lower tank. Likewise, in an electric circuit, the voltage generates a pushing force to drive the electric current. This water current corresponds to the electric current flow, while the altitude difference h between the upper and lower tanks corresponds to the voltage.

Fig. 1.5 A closed-loop water system

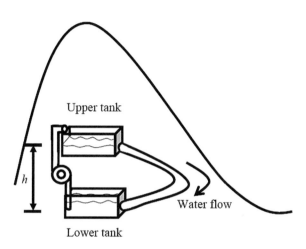

Lower tank

Resistance is a measure of the material's opposition to the flow of current and is measured in ohms (Ω). Referring back to Fig. 1.5, the water pipes have friction that inhibits the flow of water, which is similar to how resistance inhibits the flow of electrons.

Voltage, current, and resistance are closely related to each other, and these quantities change based on the type of electrical component. In order to consistently analyze these components, electrical engineers use **passive sign convention**, a method for assigning the positive and negative terminals of a component as well as the direction of current.

In Fig. 1.6, an electrical component is labeled in accordance with passive sign convention. The positive and negative terminals of the component can be arbitrarily assigned, but the current direction must be from the positive terminal to the negative terminal. It does not matter how you initially chose the positive and negative

Fig. 1.6 Voltage and current for an electrical component using passive sign convention

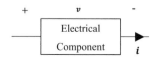

terminals, but you must be consistent for the entire analysis. Even if your answer contains a negative voltage or current, you may not have made a mistake; it just means that the terminals or the current may have been opposite of what you initially expected.

Until now, we have been using a generic representation of an electrical component, so let us look into some basic circuit elements. Current–voltage characteristic curves (**I–V curves**) represent the relationship between current and voltage for the component and can help us better understand how the component operates.

A **short circuit**, also known as a **wire**, is used to connect other components. The voltage across a wire is 0 V, while the current through a wire can be anything. The resistance is 0 Ω (Fig. 1.7).

Fig. 1.7 Representation of a wire and its I–V curve

An **open circuit** is a disconnection in the circuit. The voltage across an open circuit can be anything while the current through an open circuit is 0 A. The resistance is infinite (Fig. 1.8).

Fig. 1.8 Representation of an open circuit and its I–V curve

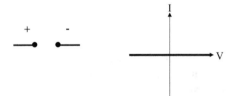

A **DC voltage source** is a circuit element that provides a fixed voltage, such as 5 V or 9 V, across it. "DC" stands for "direct current", which means that current only flows in one direction and does not change. The voltage across a DC voltage source is the voltage it is intended to provide while the current through it can be anything. The **internal resistance**, the resistance inside of a component, of a DC voltage source is 0 Ω (Fig. 1.9).

A **DC current source** is a circuit element that provides a fixed current through it. The voltage across a DC current source can be anything while the current through it is the current it is intended to provide. A DC current source's internal resistance is infinite (Fig. 1.10).

Resistors, as depicted in Fig. 1.11, are electrical components with set resistances, such as 330 Ω, 1 k Ω, and 10 k Ω. For a resistor, current is proportional to voltage, and we will further examine this relationship in Chap. 3. The resistance R in a

Fig. 1.9 Representation of a DC voltage source with voltage v and its I–V curve

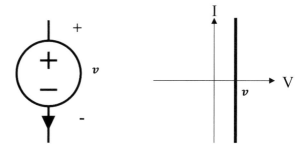

Fig. 1.10 Representation of a DC current source with current i and its I–V curve

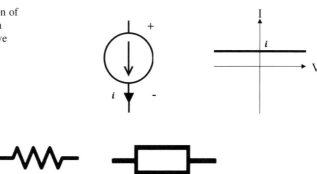

Fig. 1.11 Two representations of resistors. In this book, we will be using the one on the left, which commonly used in the USA

resistor depends on the properties of the material, geometric shape of the resistor, and sometimes temperature of the resistor. More properties of resistors will be explored in Chap. 6.

> **Notes**
> If the conductor has resistance, electric voltage is required to force the electric charges to move in one direction in a circuit, forming electric current. The circuit must be a closed loop.
>
> When using passive sign convention, make sure to stay consistent throughout the whole problem.
>
> The I–V curves here are for ideal circuit elements, which are approximations of their real-world counterparts.

Exercise Problems

Problem 1.1 Either of the following two symbols represents a DC voltage source. Here "V" is an abbreviation of "Volts." "Volt" is a unit of voltage.

Fig. P1.1

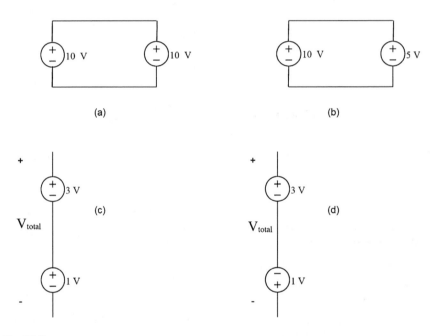

Determine whether the following configurations of voltage sources are valid or invalid. Why?

Fig. P1.2

Problem 1.2 The purpose of a voltage source in a circuit is to cause the current to flow in a circuit. The flow of the electric current can be converted into something useful to us. For example, the electric current running through a heating wire can generate heat. The electric current running through a light bulb creates light. The electric current running through an electric motor causes motion. Please comment on the circuit shown whether this circuit is useful.

Fig. P1.3

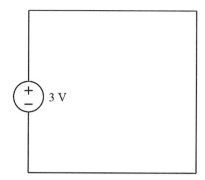

Problem 1.3 Even though we do not see them in everyday life, there are such things called "current sources." The ideal current source provides constant current, regardless the rest of the circuit. The symbol for a current source is shown below. Here "A" is an abbreviation of "Amperes." "Amperes" is a unit of current.

Fig. P1.4

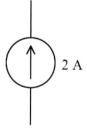

Determine whether the following configurations of current sources are valid or invalid. Why?

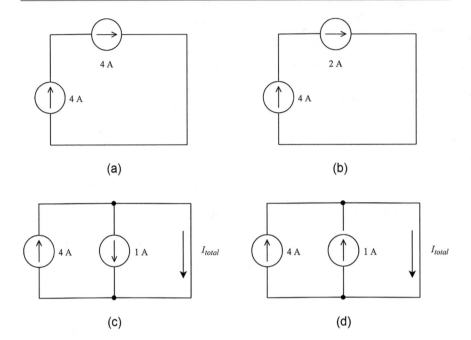

Fig. P1.5

Problem 1.4 Determine whether the following circuits are valid.

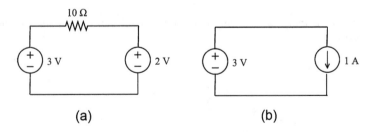

Fig. P1.6

Problem 1.5 Draw a schematic for the flashlight circuit.

Solutions to Exercise problems are given in Book Appendix.

DC Power Supply and Multimeters

2

Let us suppose you are asked to build a circuit in Fig. 2.1, then to measure the resistance of each resistor, the voltage across each resistor, and the current flowing through the circuit.

Fig. 2.1 A circuit with two 50 Ω resistors and one 6 V voltage source

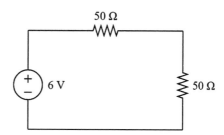

First, we will need to get two 50 Ω resistors. For the voltage source, we will be using a **DC power supply**, a device that can provide electrical power with specifications on voltage and current. To connect the circuit, we will need a breadboard and some wires. A **breadboard**, also known as a prototype board, is a board with existing internal connections that is used for building circuits. The breadboard we will be using in this example is a solderless breadboard, which contains holes for plugging in the terminals of the components.

Figure 2.2 illustrates the breadboard's internal connections, which can be thought of as wires connecting the holes. The middle two columns of the breadboard are connected horizontally, but not across the notch between these two columns. The outer two columns, also known as the **power rails**, of the breadboard are connected vertically and are typically used to connect to the power supply. By convention, the red column connects to the positive terminal, while the blue column connects to the negative terminal.

The final circuit for Fig. 2.1 is shown in Fig. 2.3. You will need to set up the DC power supply by setting the voltage to the voltage you want to supply, which is 6 V

© The Author(s), under exclusive license to Springer Nature Switzerland AG 2021
G. L. Zeng, M. Zeng, *Electric Circuits*,
https://doi.org/10.1007/978-3-030-60515-5_2

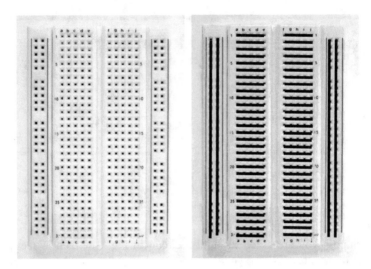

Fig. 2.2 A breadboard is shown on the left and its internal connections are shown on the right

To power supply RED (+)

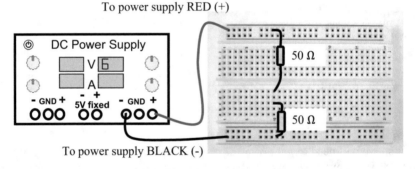

To power supply BLACK (-)

Fig. 2.3 The circuit from Fig. 2.1 is built on the board and connected to the DC power supply

in this example. For circuit protection, you should also set a limit for the current, which will vary depending on the circuit.

In some cases, we need to use more than one DC power supply in a circuit, like in Fig. 2.4. One possible way to build the circuit in Fig. 2.4 is shown in Fig. 2.5.

Your power supply panel layout may be different from the example here, so be sure to read the instructions before you connect your circuit to the power. As an example, we can build the circuit of Fig. 2.4 with a different kind of power supply as shown in Fig. 2.6.

Now that we have built a circuit, let us measure the voltage across a resistor using a **multimeter**, which is a device that can measure voltage, current, and resistance.

There are two types of multimeters: hand-held digital multimeters and desktop digital multimeters, which can be seen in Figs. 2.7 and 2.8. No matter which type of

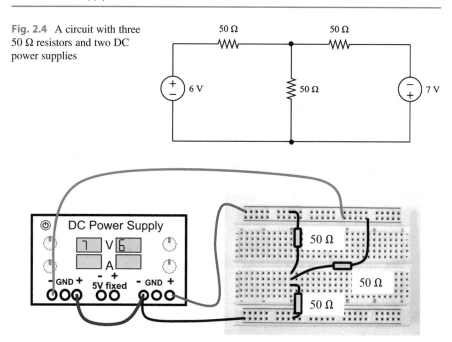

Fig. 2.4 A circuit with three 50 Ω resistors and two DC power supplies

Fig. 2.5 The circuit from Fig. 2.4 is built on the board and connected to two of the three outputs from a DC power supply

multimeter you are using, you must plug two probes into two of the proper ports of the multimeter unit in order to use it.

To measure the voltage across a resistor, you select the DC voltage measurement mode by pushing the button labeled as "DC V," connect the "Input V HI" (or "V" if the label is just "V") to one end of the resistor of interest, and connect the "LO" (or "COM" if the label is "COM" in your multimeter) to the other end of the resistor. This allows you to use the multimeter as a **voltmeter**, which measures the voltage across two points in a circuit. When you make the measurement, you must leave the power on. You can also use the voltmeter to measure the voltage across the power source, with "Input V HI" to one terminal of the power supply and "LO" to the other terminal.

Figures 2.9 and 2.10 show the setup for measuring voltage across the second resistor in the circuit from Fig. 2.4.

To measure the current, depending on your multimeter, you may need to push a button to select the DC current measurement mode, then follow the steps shown in Fig. 2.11. This allows you to use the multimeter as an **ammeter**, which measures the current through its two terminals. If you would like to measure the current through a resistor, never connect the ammeter across the resistor or across a source! You must first disconnect the circuit at a certain point. A correct connection is shown in Figs. 2.12 and 2.13. If you make a mistake, you may send too much current through the ammeter and blow the fuse.

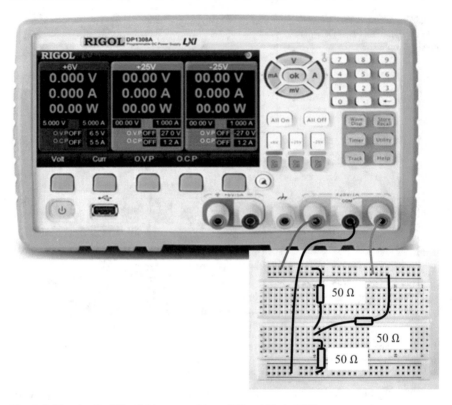

Fig. 2.6 The circuit of Fig. 2.4 is powered by a different kind of DC power supply

Fig. 2.7 A hand-held digital
multimeter

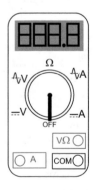

Finally, to measure the resistance of a resistor, you must remove the resistor from the circuit and measure it when the meter is at the resistance measurement mode as shown in Fig. 2.14. You may need to adjust the range to get better precision.

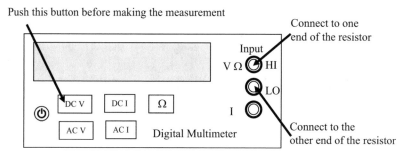

Push this button before making the measurement

Connect to one end of the resistor

Connect to the other end of the resistor

Fig. 2.8 Use of a desktop digital multimeter to measure the voltage across a resistor

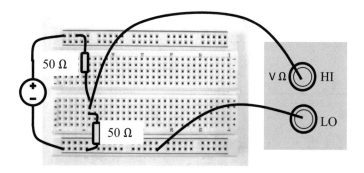

Fig. 2.9 Breadboard setup to measure voltage across a resistor

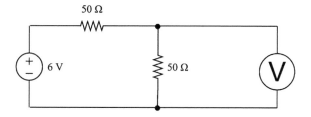

Fig. 2.10 Circuit representation of Figure 2.9, where the component with a "**V**" is the voltmeter

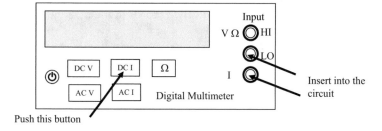

Insert into the circuit

Push this button

Fig. 2.11 Use of a desktop digital multimeter to measure current

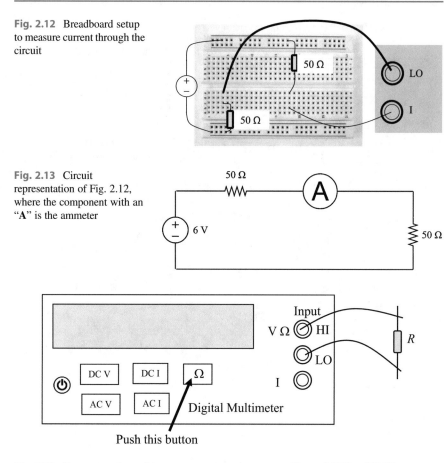

Fig. 2.12 Breadboard setup to measure current through the circuit

Fig. 2.13 Circuit representation of Fig. 2.12, where the component with an "**A**" is the ammeter

Fig. 2.14 To measure the resistance, connect a resistor across "Input Ω HI" and "LO"

Notes

To measure the *voltage* between two points, you can simply connect one probe to one point and the other probe to the other point. Be sure that the multimeter is at DC V voltage setting.

To measure the *current* at one point in the circuit, you must disconnect the circuit at that point and then insert the two probes of the multimeter there to re-connect the circuit.

To measure the *resistance* of a resistor, you need to remove the resistor from the circuit. You can disconnect at least one end of the resistor from the circuit. Never attempt to measure the resistance while the power of the circuit is on, and both ends of the resistor are still connected in the circuit.

Exercise Problems

Problem 2.1 You are given a power supply and a circuit schematic shown. Suggest three ways to connect the power supply to the 1 kΩ resistor.

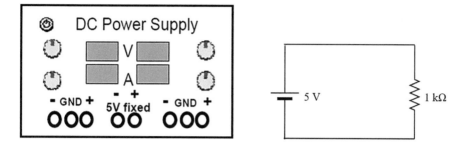

Fig. P2.1

Problem 2.2 Identify the mistakes in using a multimeter.

(a) Trying to measure the voltage across the power supply.

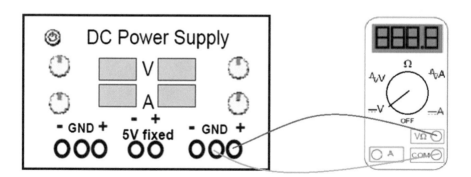

Fig. P2.2

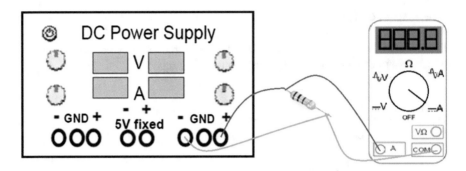

Fig. P2.3

(b) Trying to measure the current through the resistor.

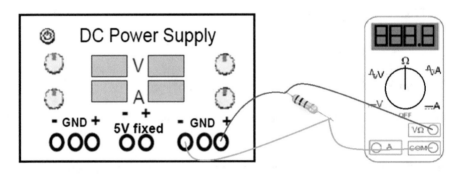

Fig. P2.4

(c) Trying to measure the resistance of the resistor.

Solutions to Exercise problems are given in Book Appendix.

Ohm's Law

3

Ohm's law is the most popular and useful law for an electrical engineer and is a must-know if you want to work with any electric circuit. Ohm's law is a relationship between the voltage v, current i, and resistance R for a resistor.

Let us revisit the water system analogy from Chap. 1 to set up an intuitive understanding of Ohm's law. In Fig. 3.1, if we move the upper water tank higher, the altitude difference between the two tanks is increased and the water will flow faster than before.

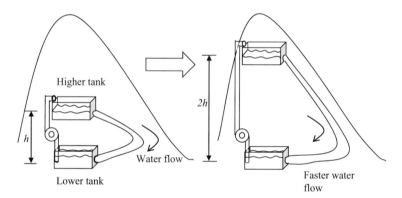

Fig. 3.1 When the altitude difference is increased, the water flow is also increased

As a side note, this water flow analogy is only a conceptual tool. In fact, this analogy is not an accurate description of an electrical system because the water flow speed varies with the altitude difference in a nonlinear relationship, while electrical current varies with voltage in a linear relationship.

If the voltage v across the resistor is doubled, then the current i through the resistor is doubled accordingly. The relationship between voltage and current is linear; in other words, the voltage v is directly proportional to the current i. Therefore, there exists such a constant R such that

$$v = Ri$$

and this constant R is the resistance. The linear relationship above is **Ohm's law**.

We apply Ohm's law to each resistor individually in a circuit, as shown in Fig. 3.2.

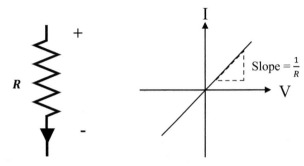

Fig. 3.2 Representation of a resistor with resistance R and Ohm's law, which describes its current–voltage relationship

Example

If the voltage across the resistor is 10 V and the current through the resistor is 50 mA, what is the resistance of the resistor?

Solution

We can calculate the resistance of the resistor using Ohm's law. We first rewrite Ohm's law to solve for R, then plug in the values provided.

$$R = \frac{v}{i} = \frac{10\text{ V}}{50\text{ mA}} = 200\ \Omega.$$

Notes

Ohm's law implies that for a given resistance R for a resistor, if you supply more voltage v across the resistor, you get more current i flowing through it. This law $v = Ri$ is the foundation of electrical engineering.

Exercise Problems

Problem 3.1 Use Ohm's law to calculate the current in the circuit.

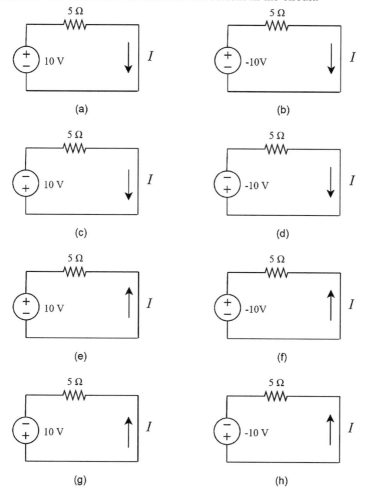

Fig. P3.1

Problem 3.2   According to the partial circuit shown, use Ohm's law to calculate the voltage across the resistor. You must use the voltage polarity and current direction specified in the figure.

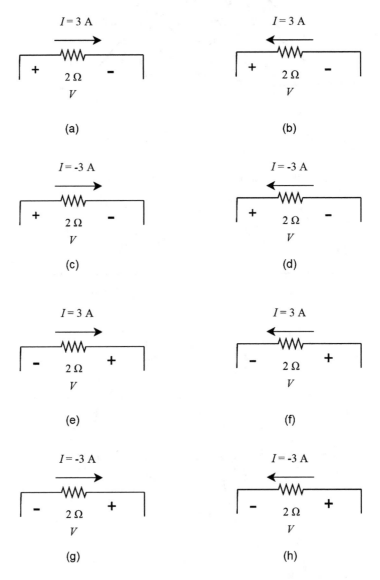

Fig. P3.2

Problem 3.3 You are given an electrical element without any labels. You connect the element with a variable voltage source. You make some voltage/current measurements as shown in the table. What most likely is this element?

V (volts)	I (amperes)
10	5
0	0
−10	−5

Fig. P3.3

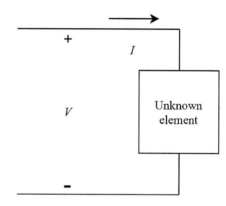

Problem 3.4 True or False?

(a) If you double the voltage across the resistor, the current through it doubles.
(b) If you double the voltage across the resistor, the current through it halves.
(c) If you halve the current through the resistor, the voltage across it doubles.
(d) If you halve the current through the resistor, the voltage across it halves.
(e) If you double the resistance of a resistor and keep the voltage across the resistor unchanged, the current through the resistor doubles.
(f) If you double the resistance of a resistor and keep the current through the resistor unchanged, the voltage across the resistor doubles.

Problem 3.5 The total human body in water is approximately 300 Ω. The electric current over 10 mA is life threatening if the current runs through the heart (10 mA = 0.01 A). How much voltage in the water can be lethal?

Solutions to Exercise problems are given in Book Appendix.

Kirchhoff's Voltage Law (KVL)

4

There are two Kirchhoff's laws, both of which are based on one concept: conservation. In this chapter, we will use hiking as an analogy to build an intuition about **Kirchhoff's voltage law** (KVL) (Fig. 4.1).

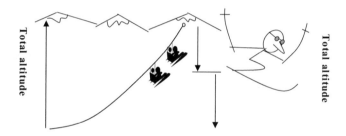

Fig. 4.1 A cartoon depiction of KVL using skiing

It is a nice weekend and there are mountains close by, so you decide to take a hike. You park your car at the trailhead parking lot and write down the altitude a_0. You can pick any trail to hike up. The requirement is that you must write down the altitude every time you take a break. After you reach your destination, you can choose any other trail to come back to your car. It is not a surprise that after you return to the parking lot, your altitude reading is the same value as what you wrote down at the beginning of your hiking trip. Otherwise, you are at the wrong parking lot!

Let us assume that your altitude records are a_0, a_1, a_2, and a_3, reflecting the path taken in Fig. 4.2. Let the altitude gain v at each hiking segment be

$$v_1 = a_1 - a_0,$$

G. L. Zeng, M. Zeng, *Electric Circuits*,
https://doi.org/10.1007/978-3-030-60515-5_4

23

Fig. 4.2 The path of a closed-loop hiking trip

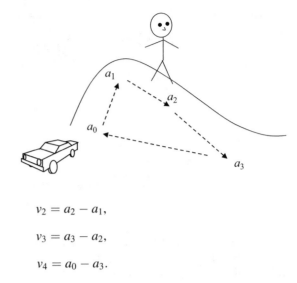

$$v_2 = a_2 - a_1,$$

$$v_3 = a_3 - a_2,$$

$$v_4 = a_0 - a_3.$$

It is important to follow the rule that the altitude gain is defined as the end point altitude minus the starting point altitude for each segment.

Some of these v values are positive, and some are negative. A positive v value implies that you hiked upwards at the hiking segment, while a negative v value implies that you hiked downwards at that segment.

Now let us sum up these v values.

$$v_1 + v_2 + v_3 + v_4 = a_1 - a_0 + a_2 - a_1 + a_3 - a_2 + a_0 - a_3 = 0.$$

The sum of the altitude gains for all segments is zero for any closed-loop hiking trip. The net altitude gain for the entire closed-loop hike is zero simply because the end point and the starting point are the same point.

The above closed-loop hiking "law" holds if we replace the "altitude gain" at each segment by the "altitude drop", which is the starting point altitude minus the end point altitude.

KVL can be applied to any electric circuit by replacing "altitude" with "electric potential" or "voltage". By KVL, in an arbitrary **loop**, or closed path, of any electric circuit, the total sum of the voltage drops across each element is zero. KVL also holds if you replace "voltage gain" with "voltage drop", but you must be consistent for the entire loop in concern. Do not mix them up.

Just like hiking where you must know the hiking direction, for a chosen electric circuit loop, you need to select a direction, which can be clockwise or counterclockwise. You can imagine that the current in this loop flows in this chosen direction. This imagined direction may be wrong, but it does not matter when you apply KVL.

Example

Figure 4.3 shows a complicated DC circuit with many closed loops, with one loop highlighted using thicker lines. Write the KVL expression for that loop with the provided direction and labeling.

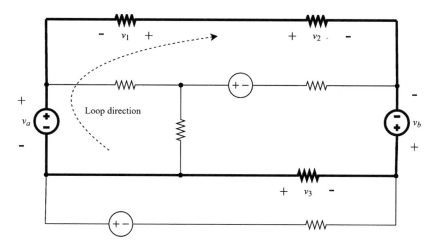

Fig. 4.3 Applying KVL to a randomly chosen loop

Solution

The loop direction has already been assigned as clockwise, and each element has already been labeled with "+" and "−" signs. We can assign each element a "direction" according to the given "+" and "−" signs, where the positive direction is from "+" to "−" following the loop direction. The negative direction is from "−" to "+" following the loop direction. For each element in the loop, we add its voltage to the existing sum of the voltages if it is in the positive direction and subtract its voltage if it is in the negative direction. Starting from v_1, the KVL equation becomes:

$$-v_1 + v_2 - v_b - v_3 - v_a = 0.$$

Instead, if we had set the positive direction to be from "−" to "+" following the loop direction, we would still have gotten a valid KVL equation. The key is to be consistent.

Some people find it easier to divide the elements into two groups: elements in the positive direction and elements in the negative direction. The resulting KVL equation can be set up by summing the voltages for each group, then setting them equal to each other.

Notes

Kirchhoff's voltage law (KVL) is a law of "what goes up must come down". In any closed loop, the voltage can increase and decrease when traversing all of the elements in the loop. The net voltage change must be zero in a loop.

This law is based on energy conservation.

Exercise Problems

Problem 4.1 Using the given voltage polarities, set up KVL equations for the following circuits:

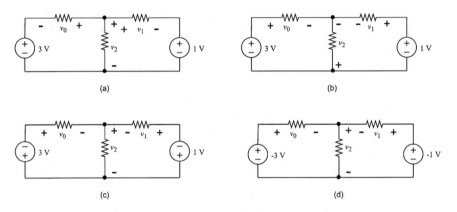

Fig. P4.1

Problem 4.2 In this problem, we will use a new current source called controlled current source, whose symbol is a diamond with an arrow inside (see the figure below). The symbol for a regular current source is a circle with an arrow inside. For example, a controlled current source is as follows:

Fig. P4.2 $2i$

Here "$2i$" indicate the value of this current source, and this value is two times the current value i, which is defined elsewhere in the circuit.

Fig. P4.3

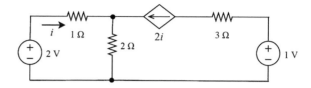

Find the current i in this circuit.

Problem 4.3 Set up the KVL equations for the following Wheatstone bridge circuit.

Fig. P4.4

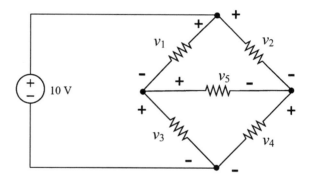

Problem 4.4 Do not simplify the circuit. Use the KVL to solve for the current i in the circuit.

Fig. P4.5

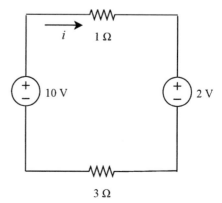

Problem 4.5 Use KVL to verify if the following circuit is valid.

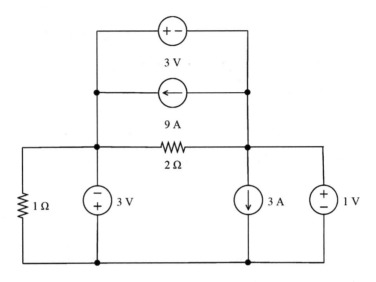

Fig. P4.6

Problem 4.6 Use KVL to verify if the following circuit is valid.

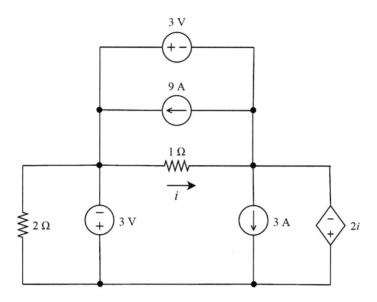

Fig. P4.7

Solutions to Exercise problems are given in Book Appendix.

Kirchhoff's Current Law (KCL)

5

Kirchhoff's current law (KCL) can be intuitively understood using the river analogy (Fig. 5.1). Rivers sometimes merge and split into other branches, like in Fig. 5.2.

Fig. 5.1 A cartoon depiction of KCL using traffic

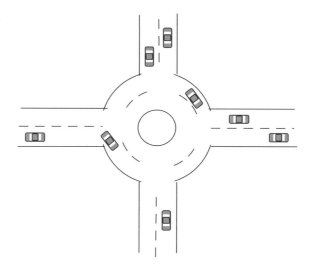

Since there is nowhere else for the water to go, the total amount of water flowing into a region is equal to the total amount of water flowing out. For the example given in Fig. 5.2, this means that

$$i_1 + i_2 = i_3 + i_4 + i_5,$$

where the water currents are labeled as i_1, i_2, ..., i_5. Likewise, by KCL, the total current entering a junction of an electric circuit is equal to the total current exiting

© The Author(s), under exclusive license to Springer Nature Switzerland AG 2021
G. L. Zeng, M. Zeng, *Electric Circuits*,
https://doi.org/10.1007/978-3-030-60515-5_5

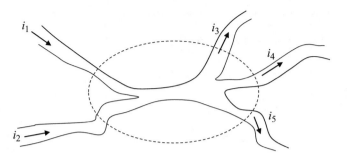

Fig. 5.2 Rivers merge and fork

that junction. A junction can either be a node or a supernode. A **node** is an uninterrupted stretch of wire that connects two or more circuit elements, while a **supernode** is a portion of the circuit that may contain multiple elements.

Example
Write the KCL expressions for the supernode and the node marked in Fig. 5.3 using i_1, i_2, i_3, i_4, and i_5.

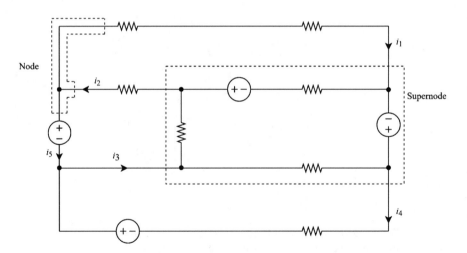

Fig. 5.3 KCL can be applied to a node or a supernode, which are marked with the dotted lines

Solution

For the supernode, we use KCL to set the total current entering the supernode equal to the total current exiting the supernode, obtaining the expression:

$$i_1 + i_3 = i_2 + i_4.$$

For the node, we use the same process to get

$$i_1 + i_5 = i_2.$$

Notes

Kirchhoff's current law (KCL) is based on the principle of conservation of electric charge. The sum of the current entering a junction is equal to the sum of the current leaving a junction.

Exercise Problems

Problem 5.1 Use KCL to verify if the following circuit is valid.

Fig. P5.1

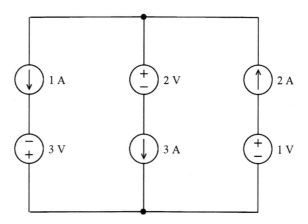

Problem 5.2 Use KCL to verify if the following circuit is valid.

Fig. P5.2

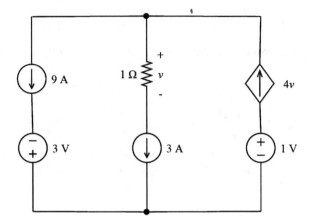

Problem 5.3 Set up the KCL equations for the following Wheatstone bridge circuit.

Fig. P5.3

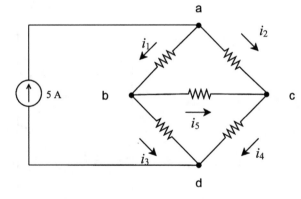

Problem 5.4 Find the current i_1 in the circuit shown in Fig. P5.4.

Fig. P5.4

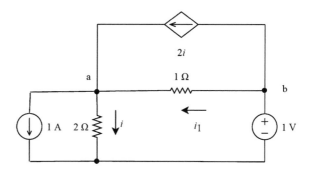

Problem 5.5 This circuit model a transistor, which has many applications such as amplifiers. Find i_b.

Fig. P5.5

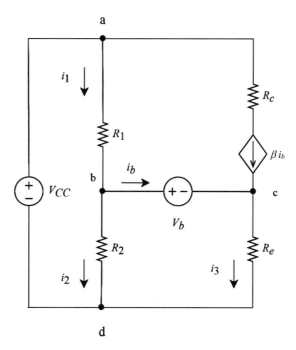

Solutions to Exercise problems are given in Book Appendix.

Resistors in Series and in Parallel

6

Being connected in parallel and in series are two common configurations to connect components, which are shown in Fig. 6.1. Components that are connected in **parallel** share the same voltage across each component, while components that are connected in **series** share the same current through each component.

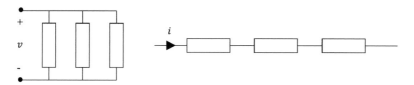

Fig. 6.1 The circuit on the left shows components connected in parallel with the same voltage v. The circuit on the right shows components connected in series with the same current i. The rectangle represents a generic circuit element

When we have resistors in parallel and in series, we can combine the resistors into one resistor with an equivalent resistance, as illustrated in Fig. 6.2.

When we combine R_1, R_2, and R_3 in parallel, we get the following expression for R_{eq}:

$$\frac{1}{R_{eq}} = \frac{1}{R_1} + \frac{1}{R_2} + \frac{1}{R_3}.$$

When we combine R_1, R_2, and R_3 in series, we get

$$R_{eq} = R_1 + R_2 + R_3.$$

The general format of the above two expressions holds for different numbers of resistors being connected, with only the number of terms being added together differing.

G. L. Zeng, M. Zeng, *Electric Circuits*,
https://doi.org/10.1007/978-3-030-60515-5_6

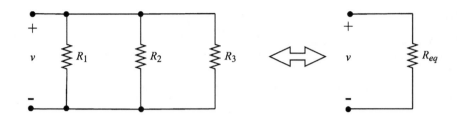

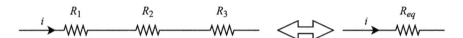

Fig. 6.2 For both resistors in parallel (top) and resistors in series (bottom), we can combine multiple resistors into one equivalent resistor R_{eq}

We can gain an intuitive understanding of the expressions for equivalent resistance by looking at the expression for the resistance R.

$$R = \rho \frac{L}{A},$$

where ρ is the resistivity of the material, L is the length, and A is the cross-sectional area, as labeled in Fig. 6.3.

Fig. 6.3 Resistance
R depends on the resistor's
length L, resistivity ρ, and area
A

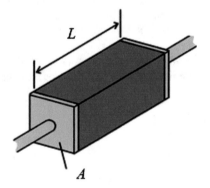

If we have three resistors with the same L, ρ, and A, then they will all have the same resistance R. Connecting these three resistors in parallel, like in Fig. 6.4, would be like forming a new resistor with area $3A$, resulting in

Fig. 6.4 The effect of three resistors in parallel

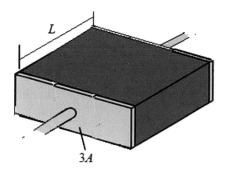

$$R_{eq} = \rho \frac{L}{3A} = \frac{R}{3},$$

which is what we expected from the expression for resistors in parallel.

Connecting those three resistors in series, as shown in Fig. 6.5, would be like forming a new resistor with length $3L$, so

$$R_{eq} = \rho \frac{3L}{A} = 3R,$$

which is what we expected from the expression for resistors in series.

Fig. 6.5 The effect of three resistors in series

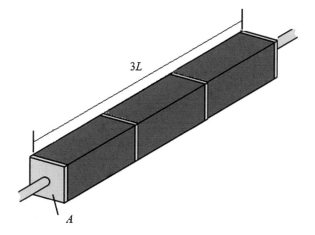

Notes

When the resistors are connected in *series*, the equivalent resistance is larger than the largest one, and the equivalent resistance is the sum of all individual resistance.

(continued)

When the resistors are connected in *parallel*, the equivalent resistance is smaller than the smallest resistance, and its reciprocal is the sum of the reciprocals of the individual resistances. The reciprocal of resistance is called **conductance**.

Exercise Problems

Problem 6.1 Ten 1 kΩ resistors are connected in series, the total resistance is

(a) 10 Ω
(b) 100 Ω
(c) 1 kΩ
(d) 10 kΩ
(e) 100 kΩ

Problem 6.2 Ten 1 kΩ resistors are connected in parallel, the total resistance is

(a) 10 Ω
(b) 100 Ω
(c) 1 kΩ
(d) 10 kΩ
(e) 100 kΩ

Problem 6.3 Two resistors R_1 and R_2 are connected in series. The total resistance is 1 kΩ.

(a) R_1 is less than 1 kΩ.
(b) R_1 is larger than 1 kΩ.
(c) R_1 is 500 Ω.
(d) R_1 and R_2 must have the same resistance.
(e) R_1 and R_2 must not have the same resistance.

Problem 6.4 Two resistors R_1 and R_2 are connected in parallel. The total resistance is 1 kΩ.

(a) R_1 is less than 1 kΩ.
(b) R_1 is larger than 1 kΩ.
(c) R_1 is 2 kΩ.
(d) R_1 and R_2 must have the same resistance.
(e) R_1 and R_2 must not have the same resistance.

Problem 6.5 Four resistors R_1, R_2, R_3, and R_4 are connected in parallel. They satisfy the relationship: $R_1 = R_2 < R_3 = R_4$. The total resistance is 1 kΩ.

(a) R_1 is less than 1 kΩ.
(b) R_1 is less than 2 kΩ.
(c) R_3 is less than 1 kΩ.
(d) R_3 is less than 2 kΩ.
(e) R_3 is less than 4 kΩ.
(f) R_1 is larger than 4 kΩ.
(g) None of the above.

Problem 6.6 Ten resistors are connected in parallel, with $R_n = n\Omega$, for $n = 1, 2, \ldots$, 10.

(a) $R_{total} = 1 + 2 + 3 + 4 + 5 + 6 + 7 + 8 + 9 + 10 = 55\ \Omega$
(b) $R_{total} = \left(\frac{1}{1} + \frac{1}{2} + \frac{1}{3} + \frac{1}{4} + \frac{1}{5} + \frac{1}{6} + \frac{1}{7} + \frac{1}{8} + \frac{1}{9} + \frac{1}{10}\right) = 2.9290\ \Omega$

(c) $R_{\text{total}} < 1\ \Omega$

(d) $R_{\text{total}} = \frac{1 \times 2 \times 3 \times 4 \times 5 \times 6 \times 7 \times 8 \times 9 \times 10}{1+2+3+4+5+6+7+8+9+10}\ \Omega$

Fig. P6.1

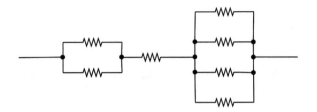

Problem 6.7 Find the total resistance for the resistor network shown in Fig. P6.1. Each resistor in the network is $1\ \Omega$.

Solutions to Exercise problems are given in Book Appendix.

Voltage Divider and Current Divider

Two circuits that are commonly used in electrical engineering are voltage dividers and current dividers. Figure 7.1 shows a typical **voltage divider** consisting of resistors in series, with the voltage across one resistor as the output voltage v_{out}. A resistor's share of the total voltage v_{in} is proportional to its resistance, so in the configuration of Fig. 7.1,

$$v_{out} = v_{in} \frac{R_2}{R_1 + R_2}.$$

If instead we were measuring the voltage across R_1 as the output voltage, we would get

$$v_{out} = v_{in} \frac{R_1}{R_1 + R_2}.$$

Fig. 7.1 An example of a typical voltage divider

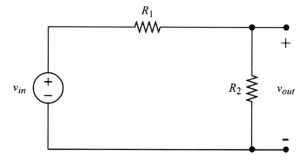

In general, the expression for v_{out} can be written as follows:

G. L. Zeng, M. Zeng, *Electric Circuits*,
https://doi.org/10.1007/978-3-030-60515-5_7

$$v_{out} = v_{in} \frac{R}{R_{total}},$$

where R is the resistance of the resistor that v_{out} is measured across, and R_{total} is the equivalent resistance of all of the resistors in series.

Figure 7.2 shows a typical **current divider** consisting of resistors in parallel, with the current through one resistor as the output current i_{out}. A resistor's share of the total current i is inversely proportional to its resistance, so for Fig. 7.2,

$$i_{out} = i \frac{R_{eq}}{R_1},$$

where R_{eq} is the equivalent resistance of all of the resistors in parallel.

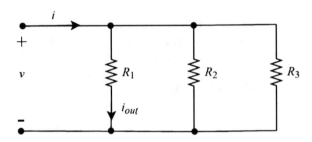

Fig. 7.2 An example of a typical current divider

The general expression for i_{out} has the same form as the expression for Fig. 7.3, but R_1 is replaced with the resistor the output current is flowing through.

Another common circuit that is based on resistors is the Wheatstone bridge. A **Wheatstone bridge** consists of four resistors and a galvanometer, as shown in Fig. 7.3. A **galvanometer**, which corresponds to the circle with a tilted needle, is a device that measures current. In Fig. 7.3, R_1, R_3, and R_4 are known, while R_2 is variable, as indicated by the arrow through the resistor symbol.

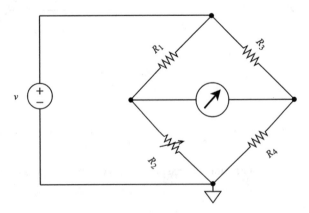

Fig. 7.3 A configuration of a Wheatstone bridge

We can redraw Fig. 7.3 in the configuration of Fig. 7.4 to clearly illustrate the bridge between the R_1–R_2 branch and the R_3–R_4 branch containing the galvanometer.

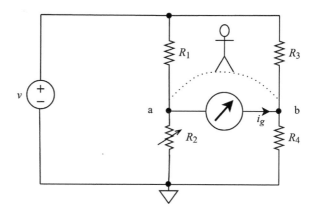

Fig. 7.4 Another way to draw the Wheatstone bridge

When the current i_g measured by the galvanometer is 0, the bridge is balanced. We can determine the conditions under which the bridge is balanced by treating each branch as a voltage divider. We can only use the voltage divider expression if the current flowing through both resistors is the same, which is the case if there is no net current flowing through the bridge. Otherwise, we cannot use the voltage divider expression for both branches.

If there is a voltage difference between nodes a and b, current will flow, so we must first set

$$v_a = v_b.$$

Using the voltage divider expression for both branches, we get

$$v_a = v\frac{R_2}{R_1 + R_2}, v_b = v\frac{R_4}{R_3 + R_4}.$$

After substituting, this simplifies to

$$\frac{R_2}{R_1 + R_2} = \frac{R_4}{R_3 + R_4},$$

$$R_2R_3 + R_2R_4 = R_1R_4 + R_2R_4,$$

and finally,

$$R_2R_3 = R_1R_4.$$

This relationship between the resistances allows us to use the Wheatstone bridge for a variety of applications, such as measuring an unknown resistance.

Notes

In a *voltage divider*, the resistor with the largest resistance gets the biggest share of the total voltage. In a *current divider*, the resistor with the largest resistance gets the smallest share of the total current.

When appropriate, using voltage dividers and current dividers to solve for voltages and currents in a circuit is much more convenient than solving equations.

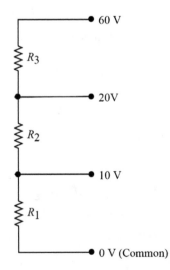

Exercise Problems

Problem 7.1 In the circuit shown in Fig. P7.1, $R_1 = 1$ kΩ. Find the values of other resistors.

Fig. P7.1

Problem 7.2 Find the voltage v in the circuit shown in Fig. P7.2. All resistors have the value of 1 Ω.

Fig. P7.2

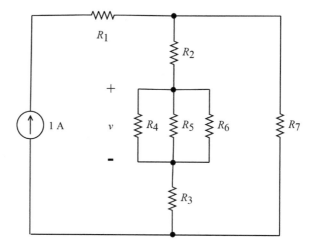

Problem 7.3 Calculate currents i_1, i_2, i_3, and i_4 in the circuit shown in Fig. P7.3.

(a) $R_1 = R_2 = R_3 = R_4 = 1\ \Omega$
(b) $R_1 = 1\ \Omega$, $R_2 = 2\ \Omega$, $R_3 = 3\ \Omega$, and $R_4 = 4\ \Omega$

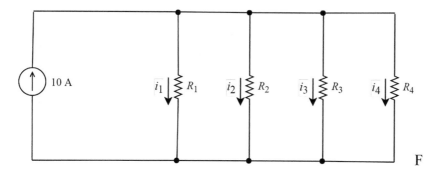

Fig. P7.3

Problem 7.4 Find i_1 and i_2 in the circuit shown in Fig. P7.4.

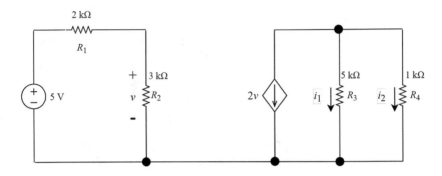

Fig. P7.4

Problem 7.5 For a voltage divider circuit shown in Fig. P7.5. R_1 is 1 Ω. The voltage across R_1 is 1 V. Is it possible to determine the source voltage v and the value of the other resistor R_2?

Fig. P7.5

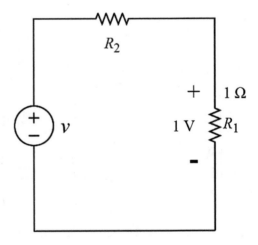

Solutions to Exercise problems are given in Book Appendix.

Node-Voltage Method

8

The **node-voltage method** is a method that solves for the voltages at each essential node in a circuit using KCL. An **essential node** is a node that connects to three or more components. We can divide the method into four steps:

1. Label each essential node and the directions of each current exiting/entering a node. Select one node as ground.
2. Write KCL equations for each essential node, excluding ground.
3. Write expressions for the currents in terms of the node voltages. For resistors, these expressions will come from Ohm's law.
4. Substitute the expressions for the currents into the KCL equations and solve.

> **Example**
> Find currents i_1, i_2, ..., and i_6 in the circuit shown in Fig. 8.1, where R_1, ..., and R_6 are known.

Solution

G. L. Zeng, M. Zeng, *Electric Circuits*,
https://doi.org/10.1007/978-3-030-60515-5_8

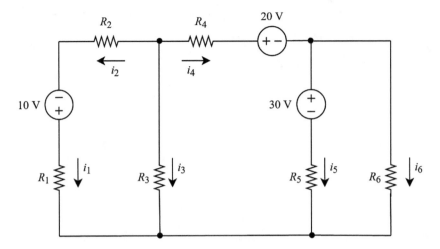

Fig. 8.1 Node-voltage method circuit

1. Figure 8.2 shows the completed first step. There are three essential nodes, which we labeled as v_1, v_2, and ground. We choose the node with the most connections to be ground, which is indicated by the symbol with three horizontal lines, to simplify our calculations.
2. The KCL equations are as follows:

$$i_2 + i_3 + i_4 = 0,$$

$$i_4 = i_5 + i_6 \text{ or } -i_4 + i_5 + i_6 = 0.$$

3. Next, we need to write expressions for each of the currents used in the KCL equations. v_1, R_2, the 10 V source, and R_1 are all in series, so we can treat this portion of the circuit like the 10 V source and R_2 swapped places to write

$$i_2 = \frac{v_1 + 10 \text{ V}}{R_1 + R_2}.$$

We directly use Ohm's law to write the expressions for the other currents.

$$i_3 = \frac{v_1}{R_3},$$

$$i_4 = \frac{v_1 - 20 \text{ V} - v_2}{R_4},$$

$$i_5 = \frac{v_2 - 30 \text{ V}}{R_5},$$

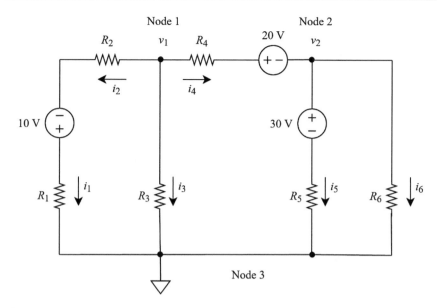

Fig. 8.2 Labeled circuit from Fig. 8.1

$$i_6 = \frac{v_2}{R_6}.$$

4. After substituting, we get the following:

$$\frac{v_1 + 10\ \text{V}}{R_1 + R_2} + \frac{v_1}{R_3} + \frac{v_1 - 20\ \text{V} - v_2}{R_4} = 0,$$

$$\frac{v_2 + 20\ \text{V} - v_1}{R_4} + \frac{v_2 - 30\ \text{V}}{R_5} + \frac{v_2}{R_6} = 0,$$

which simplifies to

$$\left(\frac{1}{R_1 + R_2} + \frac{1}{R_3} + \frac{1}{R_4}\right)v_1 + \left(\frac{-1}{R_4}\right)v_2 = \left(\frac{20\ \text{V}}{R_4} - \frac{10\ \text{V}}{R_1 + R_2}\right),$$

$$\left(\frac{-1}{R_4}\right)v_1 + \left(\frac{1}{R_4} + \frac{1}{R_5} + \frac{1}{R_6}\right)v_2 = \left(\frac{30\ \text{V}}{R_5} - \frac{20\ \text{V}}{R_4}\right).$$

From here, you can solve the system of linear equations; however, you would like. For this example, we will solve the system using matrix operations. We first rewrite the equations in matrix form:

$$
\begin{bmatrix}
\dfrac{1}{R_1 + R_2} + \dfrac{1}{R_3} + \dfrac{1}{R_4} & \dfrac{-1}{R_4} \\[3mm]
\dfrac{-1}{R_4} & \dfrac{1}{R_4} + \dfrac{1}{R_5} + \dfrac{1}{R_6}
\end{bmatrix}
\begin{bmatrix} v_1 \\ v_2 \end{bmatrix}
=
\begin{bmatrix}
\dfrac{20\text{ V}}{R_4} - \dfrac{10\text{ V}}{R_1 + R_2} \\[3mm]
\dfrac{30\text{ V}}{R_5} - \dfrac{20\text{ V}}{R_4}
\end{bmatrix}.
$$

The solution is

$$
\begin{bmatrix} v_1 \\ v_2 \end{bmatrix}
=
\begin{bmatrix}
\dfrac{1}{R_1 + R_2} + \dfrac{1}{R_3} + \dfrac{1}{R_4} & \dfrac{-1}{R_4} \\[3mm]
\dfrac{-1}{R_4} & \dfrac{1}{R_4} + \dfrac{1}{R_5} + \dfrac{1}{R_6}
\end{bmatrix}^{-1}
\begin{bmatrix}
\dfrac{20\text{ V}}{R_4} - \dfrac{10\text{ V}}{R_1 + R_2} \\[3mm]
\dfrac{30\text{ V}}{R_5} - \dfrac{20\text{ V}}{R_4}
\end{bmatrix}.
$$

5. For this particular problem, there is an additional step since we are asked to solve for the currents instead of just the node voltages. To solve for the currents, we will use the expressions we already wrote in Step 3 and plug in v_1 and v_2 from Step 4. We did not write an expression for i_1, but $i_1 = i_2$ due to KCL.

Example
Consider the circuit in Fig. 8.3. Find v_1 and v_2.

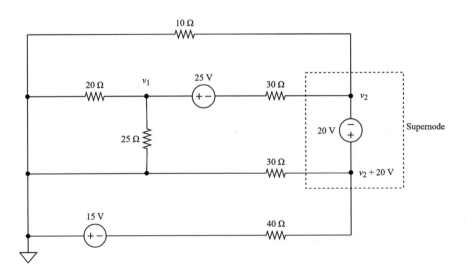

Fig. 8.3 Node-voltage method circuit with a supernode

Solution

1. The circuit in Fig. 8.3 already has one node labeled as ground and two other nodes labeled as v_1 and v_2. The reason why $v_2 + 20$ V is not labeled as v_3 is because there is only a voltage source between the two essential nodes, making it difficult to

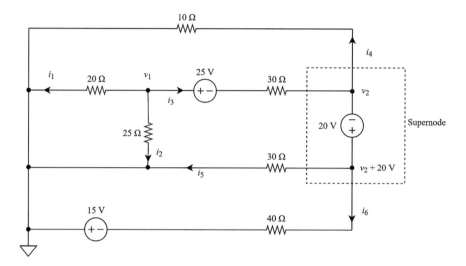

Fig. 8.4 Labeled circuit from Fig.8.3

write KCL equations, so we consolidate the two nodes into one supernode. The remainder of the labeling is done in Fig. 8.4.

2. The KCL equations are as follows:

$$i_1 + i_2 + i_3 = 0,$$

$$-i_3 + i_4 + i_5 + i_6 = 0.$$

3. Using Ohm's law, we can write

$$i_1 = \frac{v_1}{20\ \Omega},$$

$$i_2 = \frac{v_1}{25\ \Omega},$$

$$i_3 = \frac{(v_1 - 25\ \text{V}) - v_2}{30\ \Omega},$$

$$i_4 = \frac{v_2}{10\ \Omega},$$

$$i_5 = \frac{v_2 + 20 \text{ V}}{30 \ \Omega},$$

$$i_6 = \frac{(v_2 + 20 \text{ V}) - (-15 \text{ V})}{40 \ \Omega}.$$

4. After substituting, the KCL equations become

$$\frac{v_1}{20 \ \Omega} + \frac{v_1}{25 \ \Omega} + \frac{(v_1 - 25 \text{ V}) - v_2}{30 \ \Omega} = 0,$$

$$\frac{v_2 - (v_1 - 25 \text{ V})}{30 \ \Omega} + \frac{v_2}{10 \ \Omega} + \frac{v_2 + 20 \text{ V}}{30 \ \Omega} + \frac{(v_2 + 20 \text{ V}) - (-15 \text{ V})}{40 \ \Omega} = 0.$$

After simplifying, we get

$$37v_1 - 10v_2 = 250,$$

$$-10v_1 + 57.5v_2 = -712.5,$$

and finally, $v_1 \approx 3.58$ V and $v_2 \approx -11.77$ V.

This concludes the second example, but if we want to find other electrical quantities, we can solve for them using v_1 and v_2.

Notes

The *node-voltage method* provides a general guideline for analyzing circuits by setting up KCL equations. When it is difficult to set up a KCL equation, such as when there is only a voltage source between two nodes, use a supernode instead.

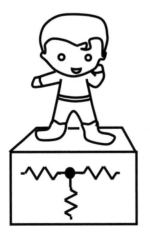 

Exercise Problems

Problem 8.1 Set up node equations for the circuit given in Fig. P8.1.

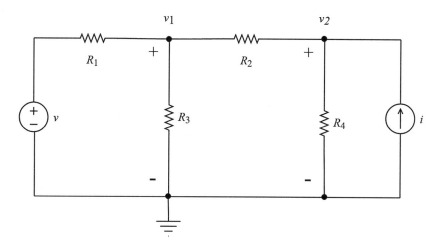

Fig. P8.1

Problem 8.2 Set up node equations for a circuit containing a controlled source.

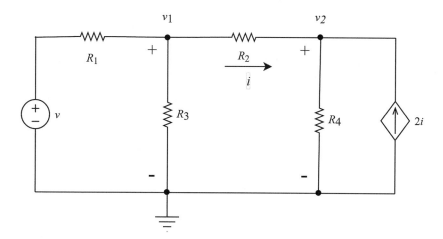

Fig. P8.2

Problem 8.3 Set up the node equations for a circuit, in which a voltage source is between the two nodes, and there are no resistors between these two nodes.

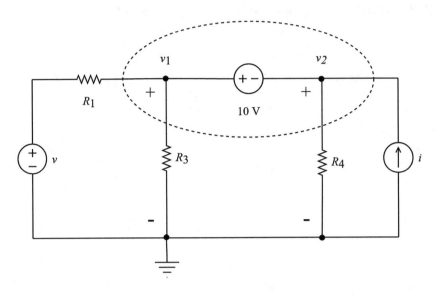

Fig. P8.3

Problem 8.4 Set up the node equations for a circuit, in which a controlled voltage source is between the two nodes, and there are no resistors between these two nodes.

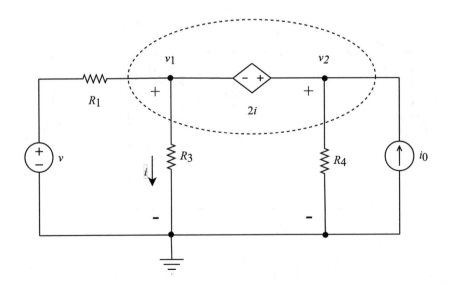

Fig. P8.4

Problem 8.5 Set up node equations for the circuit, where a voltage source is between a node and the reference node.

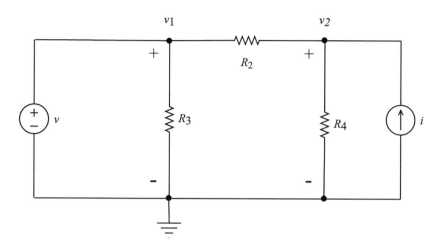

Fig. P8.5

Solutions to Exercise problems are given in Book Appendix.

Mesh-Current Method

9

The **mesh-current method** solves for the currents through each **mesh**, a loop without any inner loops, by setting up KVL equations. The mesh-current method can be split into three steps:

1. Identify each mesh and select the direction of the current (counterclockwise or clockwise) through each mesh. The direction you choose does not matter.
2. Set up a KVL equation for each mesh. Express the voltages in the KVL equations in terms of the mesh currents; for resistors, this can be done using Ohm's law.
3. Solve for the mesh currents.

Example

Find the currents i_{R1}, i_{R2}, ..., and i_{R6} in the circuit shown in Fig. 9.1.

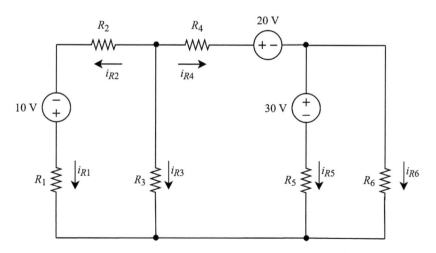

Fig. 9.1 Mesh-current method circuit

G. L. Zeng, M. Zeng, *Electric Circuits*,
https://doi.org/10.1007/978-3-030-60515-5_9

Solution

1. There are three meshes in the circuit, and we can set the direction of each mesh current as counterclockwise. Figure 9.2 contains the labeled circuit.

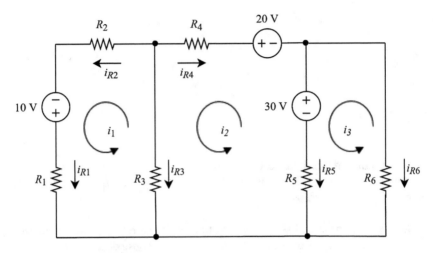

Fig. 9.2 Labeled circuit from Fig. 9.1

2. We can set up a KVL equation for each mesh current as follows. Each term in the equation represents the voltage drop across each element in the mesh.

$$R_2 i_1 - 10 \text{ V} + R_1 i_1 + R_3(i_1 - i_2) = 0,$$

$$-20 \text{ V} + R_4 i_2 + R_3(i_2 - i_1) + R_5(i_2 - i_3) - 30 \text{ V} = 0,$$

$$30 \text{ V} + R_5(i_3 - i_2) + R_6 i_3 = 0.$$

This simplifies to

$$(R_1 + R_2 + R_3)i_1 - R_3 i_2 + 0i_3 = 10 \text{ V},$$

$$-R_3 i_1 + (R_3 + R_4 + R_5)i_2 - R_5 i_3 = 50 \text{ V},$$

$$0i_1 - R_5 i_2 + (R_5 + R_6)i_3 = -30 \text{ V}.$$

3. To solve this system of equations, we can rewrite it in matrix form

$$\begin{bmatrix} R_1 + R_2 + R_3 & -R_3 & 0 \\ -R_3 & R_3 + R_4 + R_5 & -R_5 \\ 0 & -R_5 & R_5 + R_6 \end{bmatrix} \begin{bmatrix} i_1 \\ i_2 \\ i_3 \end{bmatrix} = \begin{bmatrix} 10 \\ 50 \\ -30 \end{bmatrix}$$

to get the solution:

$$\begin{bmatrix} i_1 \\ i_2 \\ i_3 \end{bmatrix} = \begin{bmatrix} R_1 + R_2 + R_3 & -R_3 & 0 \\ -R_3 & R_3 + R_4 + R_5 & -R_5 \\ 0 & -R_5 & R_5 + R_6 \end{bmatrix}^{-1} \begin{bmatrix} 10 \\ 50 \\ -30 \end{bmatrix}.$$

4. We must take an additional step to solve for the currents through the resistors. The currents through the resistors can be written as follows:

$$i_{R1} = i_{R2} = i_1,$$

$$i_{R3} = i_2 - i_1,$$

$$i_{R4} = -i_2,$$

$$i_{R5} = i_3 - i_2,$$

$$i_{R6} = -i_3.$$

Example
Find all three mesh currents in the circuit shown in Fig. 9.3.

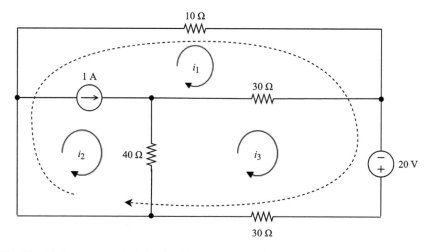

Fig. 9.3 Mesh-current method circuit with a supermesh

Solution

1. The three meshes already have the directions of their mesh currents set as clockwise. However, when there is a current source between two meshes i_1 and i_2, it becomes difficult to write KVL equations. We can address this problem by using a **supermesh**, the large loop enclosing both meshes, instead of the two meshes i_1 and i_2. We indicate the supermesh with the dotted arrow in Fig. 9.3.

2. The KVL equations for the supermesh and the mesh corresponding to i_3 are

$$10i_1 - 20 \text{ V} + 30i_3 = 0,$$

$$30i_3 + 40(i_3 - i_2) + 30(i_3 - i_1) - 20 \text{ V} = 0,$$

respectively. To solve for three unknowns, we need another equation, which we can get using the current source as a constraint.

$$i_2 - i_1 = 1 \text{ A}.$$

After simplifying, we get

$$i_1 + 0i_2 + 3i_3 = 2,$$

$$-3i_1 - 4i_2 + 10i_3 = 2,$$

$$-i_1 + i_2 + 0i_3 = 1.$$

3. Solving the system of equations gives us $i_1 \approx 0.06$ A, $i_2 \approx 1.06$ A, and $i_3 \approx 0.65$ A.

   From here, we can solve for other electrical quantities using i_1, i_2, and i_3 if we want.

Notes

The *mesh-current method* provides a general guideline for analyzing circuits by setting up KVL equations. When it is difficult to set up a KVL equation, such as when there is a current source between two meshes, use a supermesh instead.

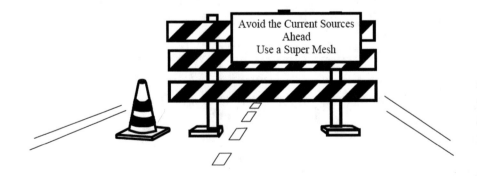

Exercise Problems

Problem 9.1 Set up the mesh equations and solve for the mesh currents.

Fig. P9.1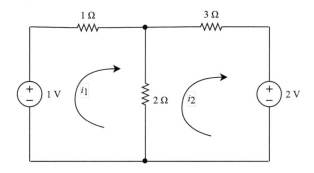

Problem 9.2 This circuit contains a voltage-controlled voltage source. Set up the mesh equations and solve for the mesh currents.

Fig. P9.2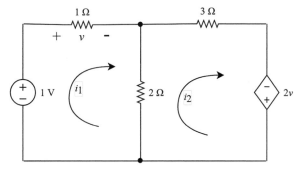

Problem 9.3 This circuit contains a current source. Set up the mesh equations and solve for the mesh currents.

Fig. P9.3

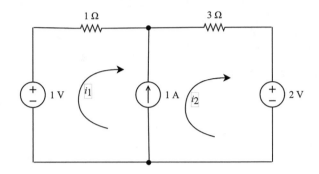

Since we do not know how to express the voltage drop across a current source, we want to avoid current source in our mesh equations. Using a super mesh can avoid the current sources. For this problem, a super mesh is indicated in Fig. P9.4 as a dotted loop.

Fig. P9.4

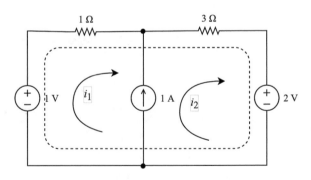

Problem 9.4 What if the current source is a controlled current source?

Fig. P9.5

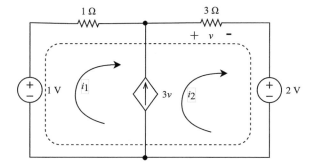

Problem 9.5 Application of the mesh equations, considering a special case of a current course in the circuit.

Fig. P9.6

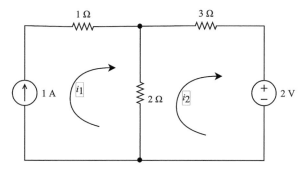

Solutions to Exercise problems are given in Book Appendix.

Circuit Simulation Software

10

Computer simulation software is crucial for circuit design because it can model a circuit's behavior, including its voltages and currents throughout the circuit, without actually building the circuit itself. The software we will be looking at in this chapter is Multisim, a popular software developed by NI for circuit analysis.

Let us go through an example to illustrate how to use Multisim from the very beginning. If you are using a Windows computer, you can run Multisim by clicking on Start → All Programs → National Instruments → Circuit Design Suite xx.x. → Multisim xx.x. This opens up a workplace, which is shown in Fig. 10.1. To create a new schematic, click on File → New → Schematic Capture. To save the schematic, click on File → Save As. To open an existing file, click on File → Open and select a file.

Fig. 10.1 Multisim workplace

© The Author(s), under exclusive license to Springer Nature Switzerland AG 2021
G. L. Zeng, M. Zeng, *Electric Circuits*,
https://doi.org/10.1007/978-3-030-60515-5_10

To place components, click on Place → Components. In the Select Component Window, click on Group to select the components that you need. Click OK to place the component on the schematic. If you click OK again, you can put another one of the same component on the schematic. If you want to rotate a component, just right click on it and select Rotate 90°.

Let us select two resistors and a DC voltage source. The resistors are in the "Basic" group, while the DC sources are in the "Sources" group, which correspond to Figs. 10.2 and 10.3, respectively. Now we have a 10-V DC voltage source and two 10 kΩ resistors, as shown in Fig. 10.4.

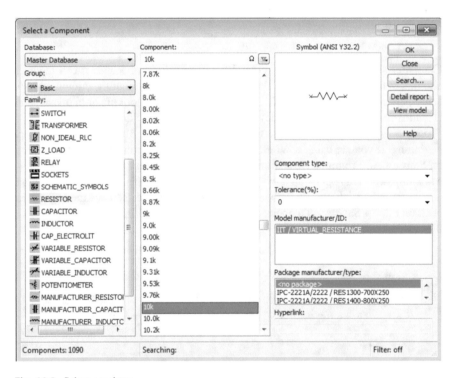

Fig. 10.2 Select a resistor

After selecting the components, we need to wire the components together. Click on Place → Wire to drag and place the wires so that the circuit looks similar to the one in Fig. 10.5.

You can change the value of a component by double-clicking that component and typing in the new value. Let us change the value of R1 to 5 kΩ. Finally, add the ground, which is in the "Sources" group because all circuits must be grounded before the circuit can be simulated. The final circuit should look something like the circuit in Fig. 10.6.

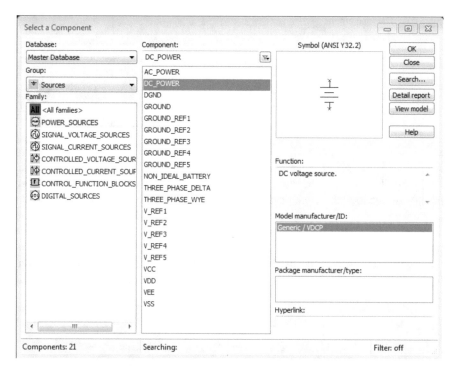

Fig. 10.3   Select a DC voltage source

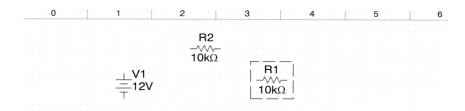

Fig. 10.4   Some components are selected to build a circuit

Fig. 10.5   The circuit after
connecting the components

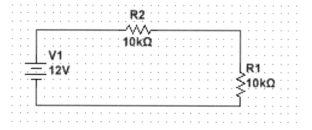

Fig. 10.6 The final circuit

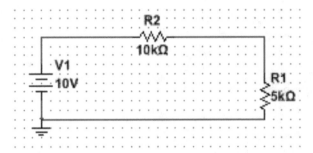

To simulate the circuit, click on Simulate → Run, or just click on the green triangle on the toolbar. Click on the red square on the toolbar to stop simulation.

Now we will take some measurements in the circuit by adding a voltmeter and an ammeter. The multimeter icon is on the vertical toolbar to the right. Put two multimeters in the workplace in the setup of Fig. 10.7 and double-click on each of the multimeters to set one as the voltmeter and the other one as the ammeter. Remember, you must disconnect the circuit and use the ammeter to reconnect the circuit in order to measure the current.

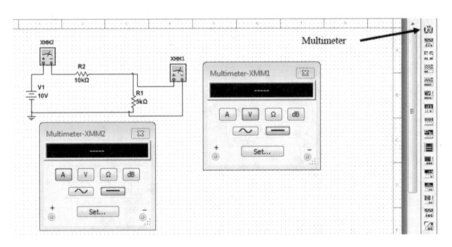

Fig. 10.7 Two multimeters are used to measure the voltage and current

After running the simulation again, you will see the measured voltage and current values in the multimeter displays shown in Fig. 10.8.

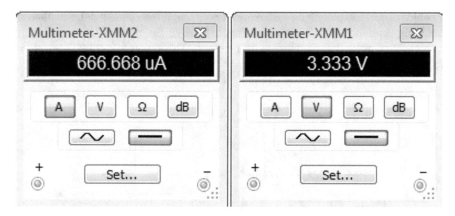

Fig. 10.8 The multimeters display the measurements

> **Notes**
> Circuit simulation software provides a means for visualizing a circuit's behavior without even building it, so it is highly recommended to simulate circuit designs before building them order to catch major flaws. Use whichever software suits the project's needs the best.

Exercise Problems

Problem 10.1 Use Multisim to simulate a circuit shown to determine the node voltage. You can choose any resistor values.

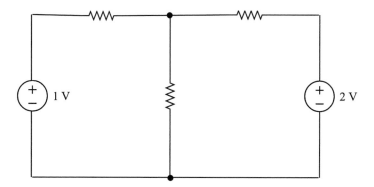

Fig. P10.1

Problem 10.2 Chapter 15 discusses operational amplifiers (Op-Amps). Simulate an inverting amplifier with a sinewave input. Change the power supply values to observe any potential clipping in the output. The op-amp circuit is given in Fig. P10.2.

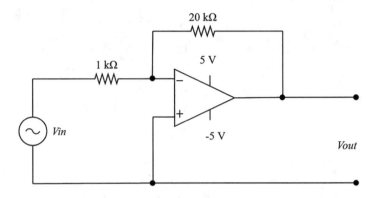

Fig. P10.2

Problem 10.3 Operational amplifier circuits are normally designed to operate from dual supplies, e.g., +9 V and −9 V. This is not always easy to achieve and therefore it is often convenient to use a single ended or single supply version of the electronic circuit design. Find a single supply op-amp circuit and use Multisim circuit.

Solutions to Exercise problems are given in Book Appendix.

Superposition

11

We see the effects of superposition in our daily life. For example, in Fig. 11.1, an audio amplification system can have multiple inputs such as voice input, music input, and so on. You can test your audio system by checking each voice input and music input separately. When all inputs come in simultaneously, your system is able to produce the mixed sound.

Fig. 11.1 A multiple-input audio system can produce mixed sound

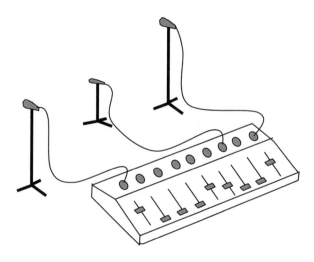

The superposition principle also works for linear input/output systems. For electric circuits, the inputs are the independent voltage and current sources, while the outputs are the voltage or current measurements at other elements. An **independent source** is a source whose output does not depend on other electrical quantities in the circuit.

Using superposition to solve for the voltage or current at a given point in the circuit can be broken up into three steps:

1. Identify all of the independent sources. In circuit diagrams, an independent source will be represented as a circle with markings on the inside to indicate whether it is a voltage or current source.
2. Choose one independent source and zero out all of the other independent sources. Zeroing out a voltage source means treating the voltage source as a short circuit while zeroing out a current source means treating the current source as an open circuit. Solve for the output at that point in the circuit. Repeat this step for all of the independent sources.
3. Add together each of the outputs obtained from Step 2 to get the total output at the given point.

During the process, do not remove any of the resistors or dependent sources. A **dependent source** is a source whose output depends on other electrical quantities in the circuit. A dependent source is also referred to as a controlled source. In circuit diagrams, a dependent source looks like a diamond with markings on the inside to indicate whether it is a voltage or current source. Voltage sources have "+" and "−" on the inside, while current sources have an arrow on the inside.

Example
In Fig. 11.2, use superposition to find the voltage v.

Fig. 11.2 Superposition circuit

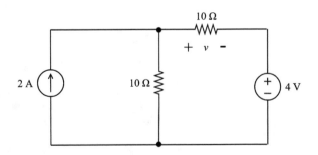

Solution
1. There are two independent sources, a 2-A current source and a 4-V voltage source.
2. Let us start with the 2-A current source, so we will first zero out the voltage source like in Fig. 11.3. The two resistors form a current divider, so the current through each resistor is 1 A. Using Ohm's law, we have

$$v_1 = (1 \text{ A})(10 \, \Omega) = 10 \text{ V}.$$

Here, v_1 is the value of v for the first case.

Fig. 11.3 The circuit after
zeroing out the voltage source

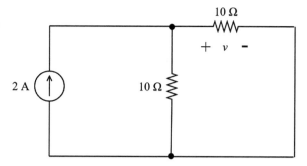

Next, we will analyze the effect of the 4 V voltage source, so we will zero out
the current source, resulting in Fig. 11.4. The two resistors form a voltage divider,
and we get

$$v_2 = -(4 \text{ V})\left(\frac{1}{2}\right) = -2 \text{ V}.$$

Here, v_2 is the value of v for the second case.

Fig. 11.4 The circuit after
zeroing out the current source

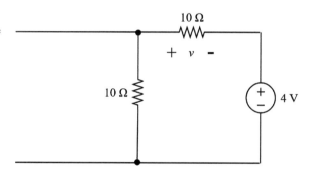

3. Combining the results from the above step gives the final result of

$$v = v_1 + v_2 = 8 \text{ V}.$$

Notes

If a circuit has multiple independent sources, analysis based on superposition
looks at one independent source at a time with all other independent sources
zeroed out. The total effect is the summation of the results from each analysis
with one source acting alone.

Do not remove any dependent source at any time, and make sure to follow
the same positive and negative labeling throughout the entire analysis.

Exercise Problems

Problem 11.1 A student uses the superposition principle to solve the voltage v. The student's answer is **wrong**. Please help this student to find the mistake. Let us start with Fig. P11.1.

Fig. P11.1

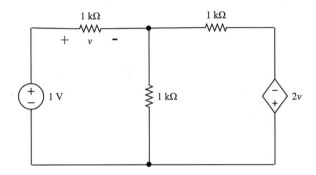

There are two sources in the circuit.

Case 1: Let us first remove the source on the left, obtaining Fig. P11.2. It can be verified that the solution is $v = 0$.

Fig. P11.2

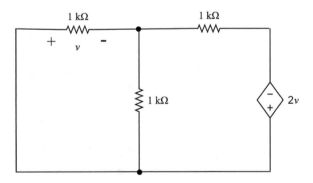

Case 2: Let us first remove the source on the right, obtaining Fig. P11.3. The left two resistors are in parallel. Therefore, these two right resistors can be combined into a 0.5-k resistor, as shown in Fig. P11.3.

This is a voltage divider, and we have $v = 2/3$ V.

Combining the results from these two steps, the final answer is $v = 2/3$ V. However, the correct answer for v is 2 V.

Fig. P11.3

Fig. P11.4

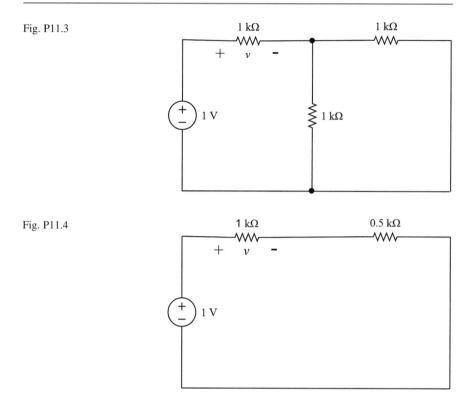

Problem 11.2 Another student tries to use the superposition principle to solve for the voltage v in a different problem. The student's answer is **wrong**. Please help this student to find the mistake. Let us start with Fig. P11.5.

Fig. P11.5

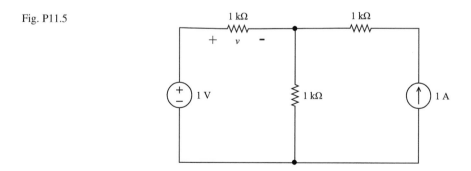

There are two independent sources in this circuit, and we will use the superposition principle to solve this problem.

Case 1: We remove the left source. The circuit becomes Fig. P11.6.

Fig. P11.6

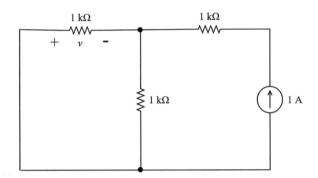

The 1 k resistor on the right has no effect in the circuit. The two resistors on the left is a current divider. Each 1kΩ resistor on the left get 0.5A of current. Using Ohm's law, $v = -500$ V.

Case 2: We remove the right source. The circuit becomes Fig. P11.7.

Fig. P11.7

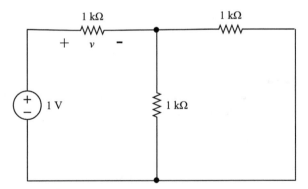

After combining the right two 1 kΩ resistors into a 0.5-kΩ resistor, we obtain a voltage divider. In this case, $v = 2/3$ V.

According to the superposition principle, we combine these two answers and obtain the final answer of

$$v = \frac{2}{3} - \frac{1}{2} = \frac{1}{6} \text{ V}.$$

However, the correct answer is

$$v = -499.5 \text{ V}.$$

What is wrong?

Problem 11.3 Use the superposition principle to solve for the voltage v. The circuit is

Fig. P11.

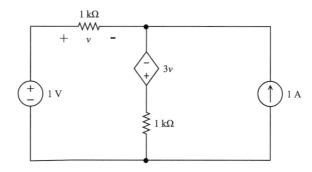

Solutions to Exercise problems are given in Book Appendix.

Thévenin and Norton Equivalent Circuits 12

For a two-terminal electric system with no source, a simple resistor can be used as an equivalent model. Very often we treat a complicated electric appliance, like a microwave oven, a vacuum cleaner, or a toaster, as a resistor. Clearly, the microwave oven is much more complicated than a simple resistor, but this resistor model is sufficient for our task.

However, many systems have sources in them and cannot be modeled as resistors. An audio system that can drive speakers is an example of such a system with sources. Therefore, we need a different way to model the circuits that make up these systems when creating equivalent circuits. An **equivalent circuit** is a simplified circuit that still exhibits the same behavior of the original circuit.

When a complex circuit only consists of sources and resistors connected linearly, we can represent it as a Thévenin equivalent circuit or a Norton equivalent circuit. A **Thévenin equivalent circuit** consolidates a circuit into a voltage source and a resistor connected in series, as illustrated in Fig. 12.1.

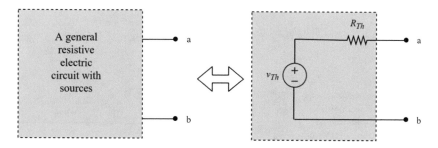

Fig. 12.1 The Thévenin equivalent circuit to model a general resistive circuit

© The Author(s), under exclusive license to Springer Nature Switzerland AG 2021   81
G. L. Zeng, M. Zeng, *Electric Circuits*,
https://doi.org/10.1007/978-3-030-60515-5_12

Finding the Thévenin equivalent circuit for a general resistive circuit requires two steps:

1. Measure v_{Th} across the terminals a and b.
2. Find R_{Th} by either measuring the short-circuit current i_{sc}, as shown in Fig. 12.2, to get $R_{Th} = \frac{v_{Th}}{i_{sc}}$ or zeroing out all of the independent sources in the circuit followed by Step 2(a) or Step 2(b).
   (a) If there are no dependent sources, combine all of the resistors between terminals a and b into one resistor with resistance R_{Th} using equivalences of resistors in series and parallel.
   (b) If there are dependent sources, add an external testing voltage or current source, as demonstrated in Fig. 12.3, then calculate R_{Th} as

$$R_{Th} = \frac{v_T}{i_T}.$$

Fig. 12.2 The Thévenin resistance R_{Th} is calculated using the short-circuit current

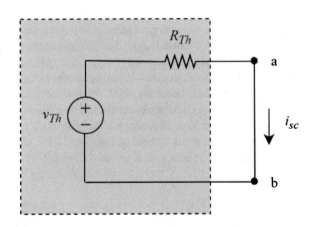

Fig. 12.3 The resistance R_{Th} is evaluated by an external testing source, with internal independent sources removed

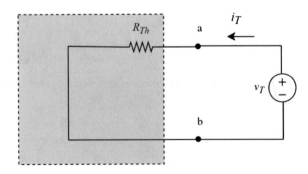

Example
Find the Thévenin equivalent circuit for the circuit in Fig. 12.4, where the
dependent current source depends on the current through the 8 Ω resistor.

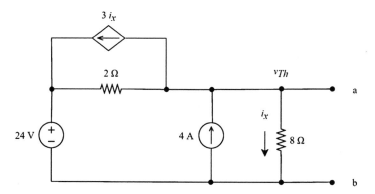

Fig. 12.4 A circuit with a dependent current source

Solution
1. We will find v_{Th} by setting up a KCL equation for node a.

$$\frac{v_{Th}}{8} + 4 + 3i_x + \frac{v_{Th} - 24}{2} = 0$$

Since

$$i_x = \frac{v_{Th}}{8},$$

$$\frac{v_{Th}}{8} + 4 + 3\left(\frac{v_{Th}}{8}\right) + \frac{v_{Th} - 24}{2} = 0,$$

and we get $v_{Th} = 8$ V.

2. To find R_{Th}, we will use the test source method with KCL. After zeroing out the
   independent sources and adding a test voltage source, we get the circuit in
   Fig. 12.5. Never remove dependent sources!
   The KCL equation for node a is

$$i_T = i_x + 3i_x + \frac{v_T}{2} = 4i_x + \frac{v_T}{2}.$$

In this case,

$$i_x = \frac{v_T}{8},$$

$$i_T = 4\left(\frac{v_T}{8}\right) + \frac{v_T}{2} = v_T.$$

Thus,

$$R_{\mathrm{Th}} = \frac{v_T}{i_T} = 1\ \Omega.$$

We can use v_{Th} and R_{Th} for the final Thévenin equivalent circuit in Fig. 12.6.

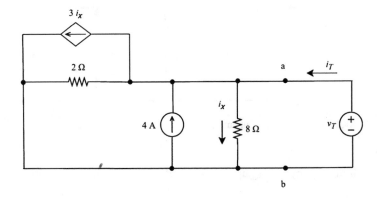

Fig. 12.5 The independent sources are removed, and an external test source is attached

Fig. 12.6 The Thévenin equivalent circuit for the circuit in Fig. 12.4

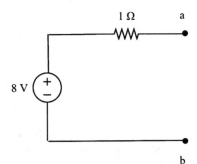

Another type of equivalent circuit is the **Norton equivalent circuit**, which consists of a current source and a resistor in parallel as shown in Fig. 12.7.

Finding the Norton equivalent circuit for a general resistive circuit requires two steps:

1. Short the terminals a and b to measure i_{Nor}.
2. Find R_{Nor} by either measuring the open-circuit voltage v_{oc} between terminals a and b to get $R_{Nor} = \frac{v_{oc}}{i_{No}}$ or zeroing out all of the independent sources in the circuit followed by Step 2(a) or Step 2(b).

   (a) If there are no dependent sources, combine all of the resistors between terminals a and b into one resistor with resistance R_{Nor} using equivalences of resistors in series and parallel.

   (b) If there are dependent sources, add an external testing voltage or current source, then calculate R_{Nor} as

$$R_{Nor} = \frac{v_T}{i_T}.$$

This process is very similar to finding the Thévenin equivalent circuit.

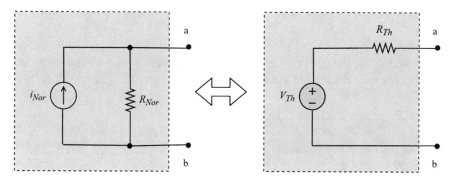

Fig. 12.7 A Norton equivalent circuit on the left and its corresponding Thévenin equivalent circuit on the right

   To understand the relationship between the two circuits in Fig. 13.1, let us find the Thévenin equivalent circuit for the Norton equivalent circuit on the left. By Ohm's law, the voltage across the terminals a and b is $i_{Nor}R_{Nor}$. R_{Th} for the Norton equivalent circuit is the resistance after removing the independent current source, so R_{Th} is R_{Nor}. Thus, for Thévenin and Norton equivalent circuits that represent the same original circuit,

$$R_{Th} = R_{Nor} \text{ and } v_{Th} = i_{Nor}R_{Nor}.$$

   To the outside world, the actual circuit and the equivalent circuit give the same results when you do circuit analysis, but the equivalent circuit is much easier to analyze. This is the purpose of using a Thévenin equivalent circuit or a Norton equivalent circuit.

   One can alternate between a Thévenin equivalent and a Norton equivalent to simplify a circuit, as illustrated in the example below.

Example
Find the Thévenin equivalent for the circuit in Fig. 12.8.

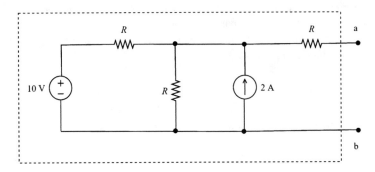

Fig. 12.8 A circuit with two sources

Solution
We will simplify the circuit from the left to the right using equivalent circuits and equivalences for resistors in parallel and in series (Figs. 12.9, 12.10, 12.11, and 12.12).

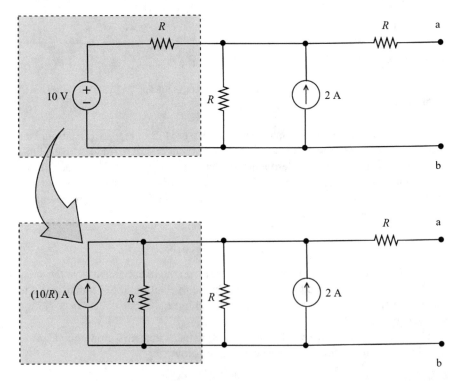

Fig. 12.9 The Thévenin circuit is converted to the Norton circuit in the shaded area. Notice that the voltage source of 10 V is converted into a current source of $(10/R)$ A

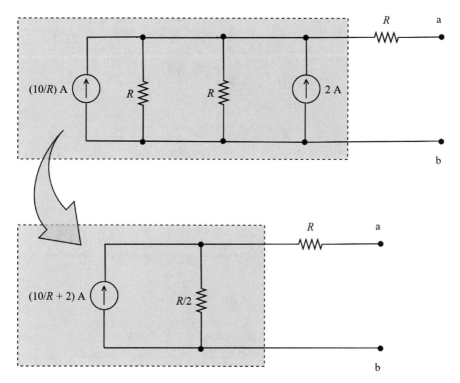

Fig. 12.10 Combining the two resistors into one and combining two current sources into one in the shaded area

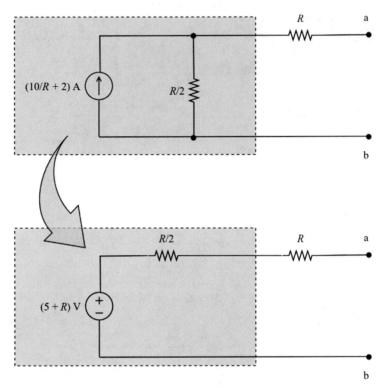

Fig. 12.11   The Norton circuit in the shaded area is converted into the Thévenin circuit. Notice that the current source of $(10/R + 2)$ A is converted into a voltage source of $(5 + R)$ V

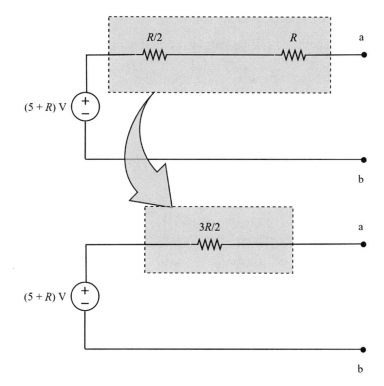

Fig. 12.12 Finally, the two resistors are combined into one resistor with a resistance $3R/2$ for the final Thévenin circuit

Notes

It is possible to simplify any linear circuit, no matter how complex, to an equivalent circuit with just a single voltage source and a resistor in series or a current source and a resistor in parallel connected to a load.

Before finding the equivalent circuit, you must first separate the load out so that the circuit has an output port.

"They're just equivalent. Neither one is the off brand."

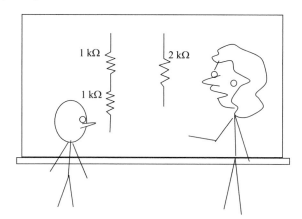

Exercise Problems

Problem 12.1 Find the Thévenin equivalent circuit of the circuit in Fig. P12.1.

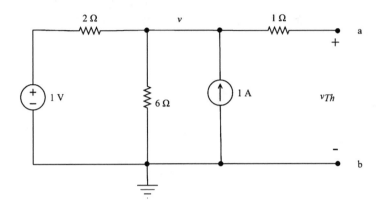

Fig. P12.1

Problem 12.2 Find the Thévenin equivalent circuit of the circuit in Fig. P12.2.

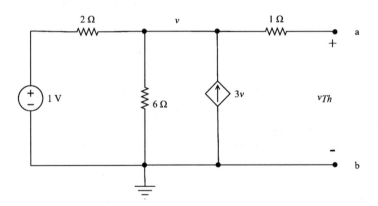

Fig. P12.2

Problem 12.3 Use the testing source method to find the Thévenin resistance R_{Th} in Problem 12.2.

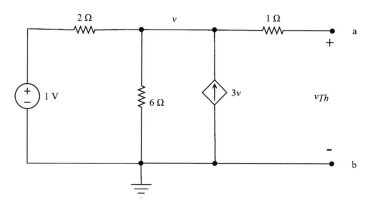

Fig. P12.3

Problem 12.4 Find the Norton equivalent circuit of the circuit in Fig. P12.4.

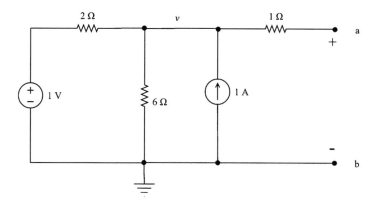

Fig. P12.4

Problem 12.5 Find the Norton equivalent circuit of the circuit in Fig. P12.5, using the step-by-step Thévenin/Norton conversion method.

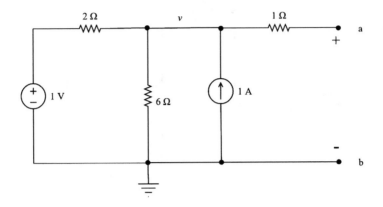

Fig. P12.5

Solutions to Exercise problems are given in Book Appendix.

Maximum Power Transfer

13

In circuits, **power** is the rate at which energy changes and is measured in watts (W). For an arbitrary component, the power P can be expressed as

$$P = IV,$$

where I is the current through the component and V is the voltage across the component, following passive sign convention. The component is dissipating power if the power is positive and supplying power if the power is negative.

For resistors specifically, $P = IV$ can be rewritten using Ohm's law into the following forms:

$$P = \frac{V^2}{R},$$

$$P = I^2 R.$$

Since the expression for power is always positive, we can see that resistors always dissipate power. With loads that are often represented as resistors, more power dissipated through the load corresponds to more perceived quantities such as louder sound from a speaker. We may want to maximize these perceived quantities, so we will need to find a way to maximize the power dissipated in the load or in other words, to find the **maximum power transfer**. This process will require using the Thévenin equivalent circuit.

Suppose that you have an audio amplifier and a speaker serving as the load, as shown in Fig. 13.1. Here, we will pretend that the audio amplifier is a DC system and everything is resistive, so the audio amplifier can be modeled by a Thévenin equivalent circuit with R_{Th} and v_{Th}. We will also treat the speaker as a resistor. What resistance should the speaker have in order to get the loudest sound?

© The Author(s), under exclusive license to Springer Nature Switzerland AG 2021
G. L. Zeng, M. Zeng, *Electric Circuits*,
https://doi.org/10.1007/978-3-030-60515-5_13

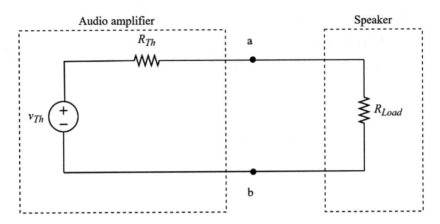

Fig. 13.1 Find the optimal resistance R_{Load} that obtains the maximum power possible

This equivalent circuit is a voltage divider. The received power by the load can be calculated as

$$p_{\mathrm{Load}} = v_{\mathrm{Load}} i_{\mathrm{Load}} = \left(v_{\mathrm{Th}} \frac{R_{\mathrm{Load}}}{R_{\mathrm{Th}} + R_{\mathrm{Load}}} \right) \times \left(\frac{v_{\mathrm{Th}}}{R_{\mathrm{Th}} + R_{\mathrm{Load}}} \right),$$

$$p_{\mathrm{Load}} = \frac{R_{\mathrm{Load}} v_{\mathrm{Th}}^2}{\left(R_{\mathrm{Th}} + R_{\mathrm{Load}} \right)^2}.$$

To find the optimal value, we can take the derivative of p_{Load} with respect to R_{Load}, and then set the derivative to zero. This procedure results in

$$R_{\mathrm{Load}} = R_{\mathrm{Th}}.$$

When R_{Load} is too small, the voltage across the load is too small, resulting in small power. When R_{Load} is too large, the current through the load is too small, also resulting in small power.

As a side note, for an AC (alternating current) circuit where the voltage and current are not constant, all the components in the equivalent circuit are complex numbers. The resistance R becomes **impedance** Z, which is the AC counterpart to resistance. The notation for R_{Load} becomes Z_{Load}, and the notation for R_{Th} becomes Z_{Th}. Maximum power transfer is achieved when the load impedance is the complex conjugate of the source impedance:

$$Z_{\mathrm{Load}} = Z_{\mathrm{Th}}^*.$$

Selecting the load impedance like this is called **impedance matching**.

Notes
The source internal resistance and the load resistance share the total power in the form of a voltage divider. When the load resistance matches the source

(continued)

internal resistance, the load gets 50% of the power that the source provides. Fifty percent of the total power is the maximum power that the load can get.

If the load resistance is larger than the matching value, the system is more efficient, but the total current is smaller than the matching current and the load gets less power.

Exercise Problems

Problem 13.1 Find the maximum power delivered to R in the circuit in Fig. P13.1 when R is set for maximum power transfer?

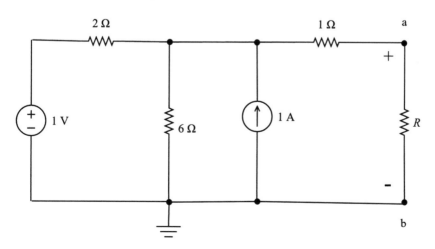

Fig. P13.1

Problem 13.2 In Problem 13.1, let $R = 2.5 \, \Omega$. What is the power provided by the 1 V voltage source? What is the power provided by the 1 A current source? In the circuit of Fig. P13.2, what percentage of the power delivered to the load $R = 2.5 \, \Omega$ by the two sources?

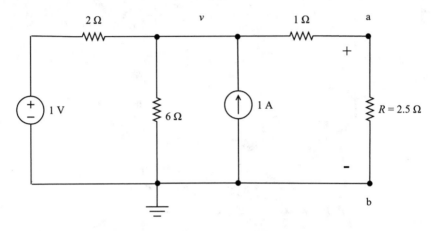

Fig. P13.2

Problem 13.3 As shown by the result of Problem 13.2, when the load resistance equals to the Thevenin resistance, the percentage of the power delivered to the load can be less than 50%. Use an example to explain this phenomenon.

Solutions to Exercise problems are given in Book Appendix.

Operational Amplifiers

14

An **operational amplifier** (op amp) is a component that amplifies the voltage difference between two input terminals and is represented by the triangle symbol in Fig. 14.1. As an amplifier, the output voltage $v_0 = A(v_p - v_n)$, where A is the amplifier gain, v_p is the voltage at the noninverting input, and v_n is the voltage at the inverting input. The output voltage cannot be greater than the positive power supply voltage or less than the negative power supply voltage. If the output voltage goes beyond these two voltages, the output will get saturated and stay at the limiting value. For an ideal amplifier, the gain A is virtually infinity, so even a slight difference between the inputs outputs a rail voltage, which refers to the power supply voltages, depending on which input is larger. This configuration is also known as a **comparator circuit**.

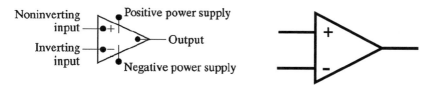

Fig. 14.1 A complete op amp symbol on the left and a simpler op-amp symbol without the positive and negative power supplies on the right

One golden rule that applies to all ideal op amps is that *virtually no current enters the input terminals*. In other words, the input impedance is very high. Op amps also have very low output impedance, which means that the output voltage does not change much even with different loads. However, there is more to consider for op amps in the commonly used configurations of negative feedback. An op amp in **negative feedback** feeds the output voltage back to the inverting input terminal, with an example being the left circuit in Fig. 14.2. Be careful not to connect the circuit like the right circuit in Fig. 14.2 because the circuit will instead be in positive feedback, resulting in an unstable circuit.

© The Author(s), under exclusive license to Springer Nature Switzerland AG 2021   97
G. L. Zeng, M. Zeng, *Electric Circuits*,
https://doi.org/10.1007/978-3-030-60515-5_14

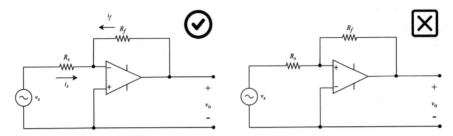

Fig. 14.2 A special type of negative feedback configuration called an inverting amplifier on the left and an unstable circuit on the right

This brings us to the golden rule that only applies to op amps in negative feedback: *for negative feedback, there is virtually no voltage difference between the noninverting input and the inverting input.* This is because the negative feedback configuration reduces the voltage difference between the two input terminals to zero while stabilizing the output. Using the golden rules appropriately makes circuit analysis much simpler.

Let us analyze the left circuit in Fig. 14.2. Since no current enters the op amp at the input terminals,

$$i_f = -i_s.$$

Since there is no voltage difference between the input terminals and v_+ is connected to ground,

$$v_- = v_+ = 0 \text{ V}.$$

Ohm's law yields

$$i_s = \frac{v_s}{R_s} \text{ and } i_f = \frac{v_o}{R_f}.$$

Thus,

$$\frac{v_s}{R_s} = -\frac{v_o}{R_f},$$

that is,

$$v_0 = -\frac{R_f}{R_s} v_s.$$

The expression indicates that the input signal v_s is inverted and amplified by a factor of $\frac{R_f}{R_s}$ to get the output signal v_0, which is why this circuit is called an **inverting amplifier**.

Let us take a look at another example of negative feedback, the **non-inverting amplifier** in Fig. 14.3.

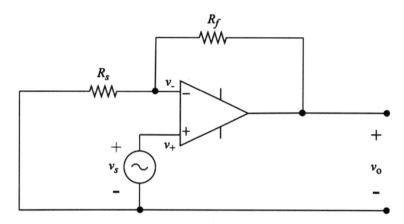

Fig. 14.3 A basic non-inverting amplifier

Since the current flowing into the op-amp is zero, the two resistors form a voltage divider. Since there is virtually no voltage drop between the input terminals, we have

$$v_- = v_+ = v_s.$$

The voltage divider relationship yields

$$v_s = v_o \frac{R_s}{R_f + R_s}.$$

Thus,

$$v_o = v_s \frac{R_f + R_s}{R_s} = v_s \left(1 + \frac{R_f}{R_s} \right).$$

The expression contains no minus sign and only amplifies v_s by $\left(1 + \frac{R_f}{R_s}\right)$, hence why it is called a non-inverting amplifier.

Another circuit with an op amp in negative feedback is the **voltage follower** in Fig. 14.4.

There is no voltage drop between v_+ and v_- and v_o is directly connected to v_-, so

$$v_s = v_- = v_o.$$

Fig. 14.4 A voltage follower

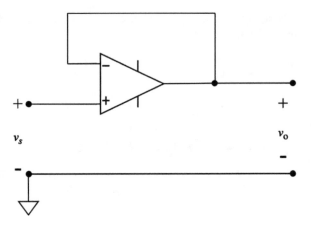

The output signal follows the input signal. Since op amps have a high input impedance and low output impedance, voltage followers are typically used as buffers to deliver more power to the load connected to the output.

Let us imagine an audio amplifier circuit as a Thévenin equivalent with a high output impedance, say 80 kΩ, connected to a speaker as a low impedance load, say 8 Ω. Directly connecting the load results in the load only getting 0.01% of v_{Th} because the circuit acts like a voltage divider. However, if we insert a voltage follower between the load and the audio amplifier as shown in Fig. 14.5, the load can get 100% of v_{Th} across it.

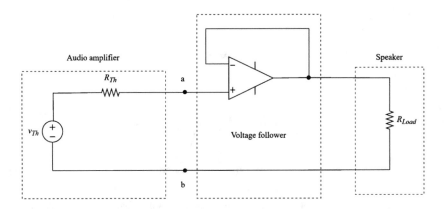

Fig. 14.5 A voltage follower is used to deliver to power to the load

Op amp circuits are not constrained to those discussed above, with an example being Fig. 14.6. Using superposition for the circuit in Fig. 14.6, we can see the following:

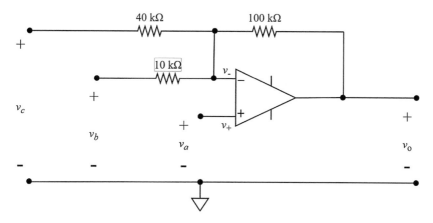

Fig. 14.6 This circuit calculates $v_o = 13.5v_a - 10v_b - 2.5v_c$

- If only v_c is available, this is an inverting amplifier and the output is

$$-\frac{100}{40}v_c.$$

- If only v_b is available, this is also an inverting amplifier and the output is

$$-\frac{100}{10}v_b.$$

- If only v_a is available, this is a non-inverting amplifier and the output is

$$\left(1 + \frac{100}{40|10}\right)v_a = \left(1 + \frac{100}{8}\right)v_a = \frac{108}{8}v_a.$$

Therefore,

$$v_o = \frac{108}{8}v_a - \frac{100}{10}v_b - \frac{100}{40}v_c = 13.5v_a - 10v_b - 2.5v_c.$$

When the output of the op amp is connected with a load R_{load}, do not apply KCL at the op amp output terminal because we do not know how much current is flowing out from the op amp. For an ideal op amp, the output voltage v_o is independent from the load and a range of currents can flow out from the output. Remember not to short the output to the ground!

Notes

Op amps are popular as building blocks in analog circuits since they can perform many mathematical operations. They are also useful for their high input impedance and low output impedance. The ideal output impedance is zero, so 100% of the output power can be delivered to the load.

For all ideal op amps, since the input impedance is so high, the current flowing into the non-inverting (+) pin and the inverting (−) pin are negligibly small. It is only valid to assume that the difference in voltage between the non-inverting (+) pin and the inverting (−) pin is negligibly small when the op amp is in negative feedback.

If the op amp is in positive feedback, we cannot assume that that the difference in voltage between the non-inverting (+) pin and the inverting (−) pin is negligibly small. A positive feedback op amp circuit is not totally useless; it can be used to construct an oscillator.

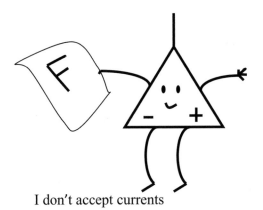

I don't accept currents

I work well with negative feedback

Exercise Problems

Problem 14.1 Express the output voltage v_{out} in terms of the inputs v_1 and v_2.

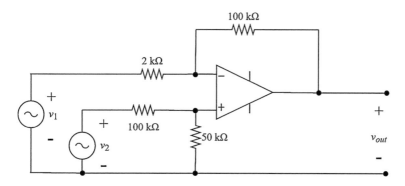

Fig. P14.1

Problem 14.2 Consider a current source as the inverting input. Find the current running into the output terminal.

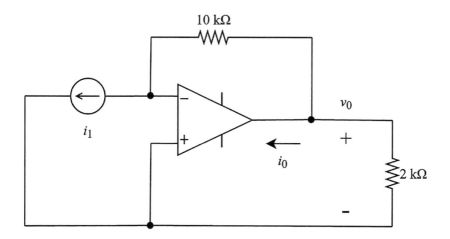

Fig. P14.2

Problem 14.3 The circuit shown in Fig. P14.3 can be thought of a current source. Find the range of the load R_L, in which the current in R_L is a constant. What is the value of this constant current?

Fig. P14.3

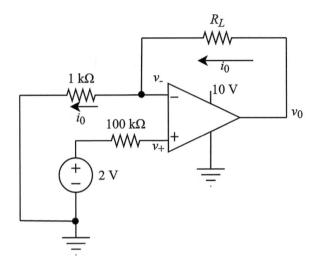

Solutions to Exercise problems are given in Book Appendix.

Inductors

15

Fig. 15.1 A photo of inductors and the symbol for inductors

An **inductor** is a coiled wire, but it does not behave like a straight wire at all. A photo of three inductors and the inductor symbol are shown in Fig. 15.1. An inductor is characterized by its **inductance** L, giving rise to a new relationship between its voltage and current:

$$v = L\frac{di}{dt}.$$

Inductors have some special properties that resistors do not have, such as how inductors can store energy in the form of a magnetic field. To obtain an expression for the energy stored in an inductor, we begin with the following general expression relating power (p) to energy (w):

$$p = \frac{dw}{dt}.$$

Power can also be related to voltage and current as

$$p = vi = \left(L\frac{di}{dt}\right)i.$$

After integrating the above expressions with respect to time, we have

$$w = \frac{1}{2}Li^2.$$

Another property is that an inductor does not allow sudden changes in current. In fact, there exists a quantity called the **time constant** which governs how slowly its current changes. For inductors, the time constant τ is equal to $\frac{L}{R}$. In response to a switch action at $t = 0$, the current through the inductor as a function of time can be found using three important values:

1. The initial current value $i(0)$.
2. The final current value $i(\infty)$.
3. The time constant $\tau = \frac{L}{R}$.

When calculating the time constant, we should use the R that is connected to the inductor after the switch action. The general mathematical expression for $i(t)$ is given as

$$i(t) = i(\infty) + [i(0) - i(\infty)]e^{-\frac{t}{\tau}}, \quad \text{for} \quad t \geq 0.$$

The following example illustrates how an inductor reacts to a switch action.

Example
Consider the circuit in Fig. 15.2. The inductor is ideal and there is no resistance in the inductor. The switch has been closed for a long time. At $t = 0$, the switch suddenly opens. What happens to the current in the inductor?

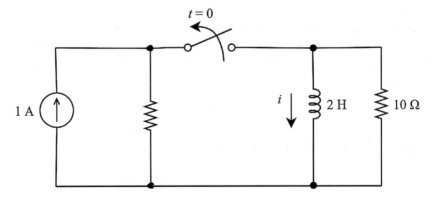

Fig. 15.2 A circuit with an inductor

Solution

While the switch is closed, the current source sends all 1 A current through the inductor because there is no resistance through the inductor.

At $t = 0$, the current source stops providing current to the inductor. The current in the inductor will slowly change to a new value.

What is the new value? There is no source to the right of the switch and there is a resistor connected to the inductor, so the resistor consumes electric energy stored in the inductor and converts it into heat while no energy is being added to the inductor. After a long transition time, the current through the inductor will approach zero, as shown in Fig. 15.3.

Fig. 15.3 The curve of the current in the inductor vs. time

In this example, $i(0)$ is 1 A and $i(\infty)$ is 0 A. The R that is connected to the inductor after the switch action is 10 Ω, so the time constant $\tau = \frac{L}{R} = \frac{2\,H}{10\,\Omega} = 0.2$ s.

Plugging these values into the general expression for $i(t)$, we get

$$i(t) = e^{-5t}\ A, \text{ for } t \geq 0.$$

We can also determine the inductance L from a typical inductor structure shown in Fig. 15.4.

Fig. 15.4 A typical inductor

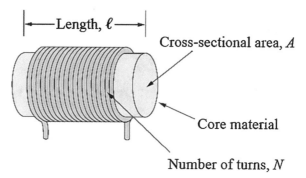

$$L = \frac{N^2 \mu A}{l},$$

where N = the number of turns, l = length, A = cross-sectional area, and μ = permeability of the core.

The equivalent inductance for inductors in series, like in Fig. 15.5, can be calculated as

$$L_{eq} = L_1 + L_2 + \ldots + L_N.$$

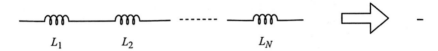

L_1 $\quad\quad$ L_2 $\quad\quad\quad\quad$ L_N

Fig. 15.5 Inductors in series

Similarly, the equivalent inductance for inductors in parallel, like in Fig. 15.6, can be calculated as

$$\frac{1}{L_{eq}} = \frac{1}{L_1} + \frac{1}{L_2} + \ldots + \frac{1}{L_N}.$$

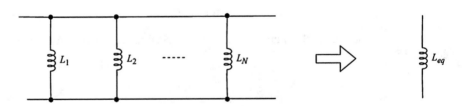

Fig. 15.6 Inductors in parallel

Notes
The voltage–current relationship of an inductor includes a derivative rather than following Ohm's law. At very low frequencies, the inductors behave almost like a short circuit while at very high frequencies, they behave almost like an open circuit. If there is a resistor connected to the inductor, the current through the inductor cannot change suddenly.

(continued)

Generally speaking, any conductor can will have some inductive properties and can be viewed as an inductor, but typical inductors are made from a cylindrical coil of conducting wire for increased inductance.

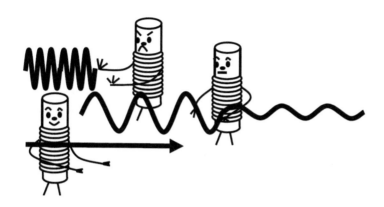

Exercise Problems

Problem 15.1 The switch closes at $t = 0$. Find the inductor current i_L as function of time.

Fig. P15.1

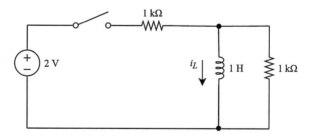

Problem 15.2 We use the same circuit as in Problem 15.1. We assume that the switch has been closed for a long time. The switch opens at $t = 0$. Find the inductor current i_L as function of time.

Fig. P15.2

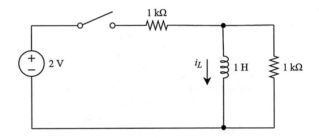

Problem 15.3 The switch in the circuit in Fig. P15.3 has been closed for a long time before opening at $t = 0$. Find the inductor's current i_L and the inductor's voltage v_L for $t \geq 0$.

Fig. P15.3

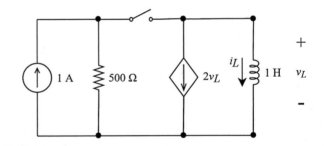

Solutions to Exercise problems are given in Book Appendix.

Capacitors

Fig. 16.1 A photo of capacitors and the symbol to represent a capacitor

A **capacitor**, as shown in Fig. 16.1, consists of two conducting layers separated by dielectric material, or in other words, an insulator. As a result, no DC current can pass through a capacitor. When there is voltage across a capacitor, an electric field is generated, causing positive charge to build up on one plate of the capacitor and negative charge to build up on the other plate.

A capacitor can be characterized by a quantity called **capacitance** C with the following relationship between its voltage and current:

$$v = \frac{1}{C} \int i(t) \mathrm{d}t.$$

Capacitors have some special properties that resistors do not have. For example, the voltage across a capacitor cannot suddenly change. Capacitors also have a time constant $\tau = RC$ which governs how slowly its voltage changes.

We can also write an expression for the capacitor voltage curve in response to a switch action, which is similar to the inductor current curve. Assuming that there is a switch action at time $t = 0$, the capacitor voltage curve is determined by three important values:

1. The initial voltage value $v(0)$.
2. The final voltage value $v(\infty)$.
3. The time constant $\tau = RC$.

G. L. Zeng, M. Zeng, *Electric Circuits*,
https://doi.org/10.1007/978-3-030-60515-5_16

We should use the R that is connected to the capacitor after the switch action to calculate the time constant τ. The general mathematical expression $v(t)$ for the voltage across a capacitor is given as

$$v(t) = v(\infty) + [v(0) - v(\infty)]e^{-\frac{t}{\tau}}, \text{ for } t \geq 0.$$

Example
Consider the circuit in Fig. 16.2. The capacitor is ideal and there is no DC current leaking through it. The switch has been closed for a long time. At $t = 0$, the switch suddenly opens. What happens to the voltages across each of these two capacitors?

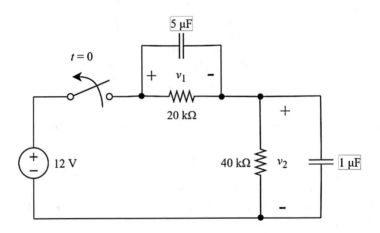

Fig. 16.2 A circuit with two capacitors

Solution
When the switch is closed for a long time, DC current does not flow through the capacitors, so the capacitors at this time can be treated as open circuits. Therefore, we can ignore the capacitors when finding the initial values. The resulting circuit resembles a voltage divider with a 12 V source, giving us

$$v_1(0) = 12 \times \frac{20}{20 + 40} = 4 \text{ V},$$

$$v_2(0) = 12 \times \frac{40}{20 + 40} = 8 \text{ V}.$$

After the switch is open, no new energy will be added to the capacitors, so the energy stored in the capacitors will be consumed by the resistors, which will convert the energy to heat. Eventually, the capacitor voltages will discharge to zero.

$$v_1(\infty) = v_2(\infty) = 0.$$

There are two time constants. The horizontal capacitor/resistor pair has a time constant of

$$\tau_1 = R_1 C_1 = (20 \text{ k}\Omega)(5 \text{ μF}) = 100 \text{ ms} = 0.1 \text{ s}.$$

The vertical capacitor/resistor pair has a time constant of

$$\tau_2 = R_2 C_2 = (40 \text{ k}\Omega)(1 \text{ μF}) = 40 \text{ ms} = 0.04 \text{ s}.$$

Thus,

$$v_1(t) = 4e^{-\frac{t}{0.1}} = 4e^{-10t} \text{ V, for } t \geq 0,$$

$$v_2(t) = 8e^{-\frac{t}{0.04}} = 8e^{-25t} \text{ V, for } t \geq 0.$$

The capacitance C can be described by the following expression involving physical characteristics of a capacitor:

$$C = \varepsilon \frac{A}{d},$$

where

C = capacitance [F],
ε = dielectric constant [N/A^2],
A = overlapping area [m^2],
d = gap [m] (Fig. 16.3).

Equivalent capacitance for capacitors in series can be calculated as (see Fig. 16.4) follows:

$$\frac{1}{C_{eq}} = \frac{1}{C_1} + \frac{1}{C_2} + \ldots + \frac{1}{C_N}.$$

Equivalent capacitance for capacitors in parallel is given as (see Fig. 16.5) follows:

$$C_{eq} = C_1 + C_2 + \ldots + C_N.$$

Fig. 16.3 A typical capacitor
consists of two conductors
and dielectric material in
between

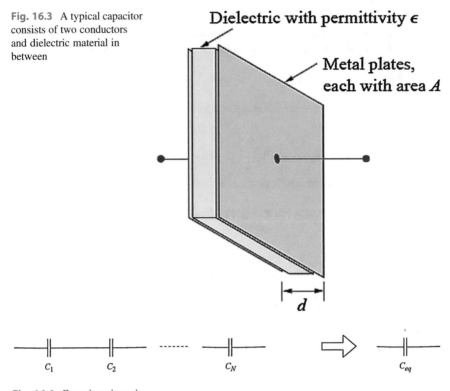

Fig. 16.4 Capacitors in series

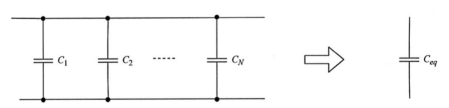

Fig. 16.5 Capacitors in parallel

Note that the expressions for series and parallel connections are opposite for capacitors and inductors. The expression for series capacitors resembles the expression for parallel inductors. The expression for parallel capacitors resembles the expression for series inductors.

The capacitor also can store energy (w). **Power** (p) is related to **energy** by

$$p = \frac{dw}{dt}.$$

Power can also related to voltage and current as

$$p = vi = v\left(C\frac{dv}{dt}\right).$$

After integration of the above expressions over time, we have $w = \frac{1}{2}Cv^2$. This voltage-generated energy is in the form of electric field.

- An inductor acts almost like a short circuit at DC (and very low frequency) and open circuit at high-frequency AC.

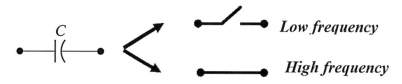

- A capacitor acts almost like an open circuit at DC (and very low frequency) and short circuit at high-frequency AC.

Notes
The capacitors do not follow Ohm's law, while the voltage and current are related by a derivative expression.

At very low frequencies, the capacitors behave almost like an open circuit. At very high frequencies, they behave almost like a short circuit.

The voltage across the capacitors cannot change suddenly.

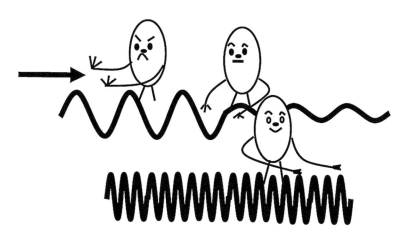

Exercise Problems

Problem 16.1 The switch closes at $t = 0$. Find the inductor current i_L as function of time.

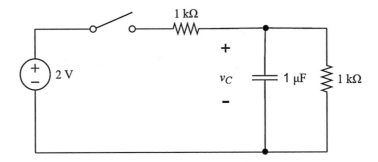

Fig. P16.1

Problem 16.2 We use the same circuit as in Problem 16.1. We assume that the switch has been closed for a long time. The switch opens at $t = 0$. Find the capacitor's voltage v_C as function of time.

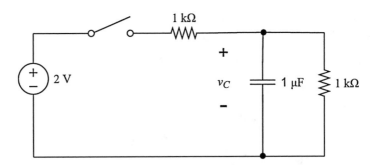

Fig. P16.2

Problem 16.3 The switch in the circuit in Fig. P16.3 has been closed for a long time before opening at $t = 0$. Find the capacitor's voltage v_C and the capacitor's current i_C for $t \geq 0$.

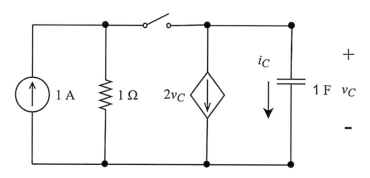

Fig. P16.3

Solutions to Exercise problems are given in Book Appendix.

Analysis of a Circuit by Solving Differential Equations

17

When a circuit contains capacitors and/or inductors, we must use derivatives to relate voltages and currents.

For a capacitor:

$$i_C(t) = C\frac{dv_C(t)}{dt}.$$

For an inductor:

$$v_L(t) = L\frac{di_L(t)}{dt}.$$

Fig. 17.1 An RC circuit

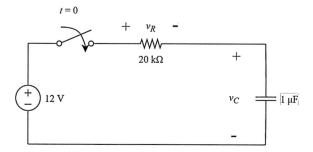

Example
Find the voltage v_R after $t = 0$.

© The Author(s), under exclusive license to Springer Nature Switzerland AG 2021
G. L. Zeng, M. Zeng, *Electric Circuits*,
https://doi.org/10.1007/978-3-030-60515-5_17

Solution

After $t = 0$, we set up a KVL equation for the RC circuit shown in Fig. 17.1 as

$$v_R + v_C = 12,$$

$$i_C R + v_C = 12,$$

$$RC \frac{dv_C(t)}{dt} + v_C = 12,$$

$$\frac{dv_C(t)}{dt} = -\frac{1}{RC}(v_C - 12),$$

$$\frac{dv_C(t)}{v_C - 12} = -\frac{1}{RC} dt,$$

$$\frac{d[v_C(t) - 12]}{[v_C - 12]} = -\frac{1}{RC} dt.$$

We will integrate both sides. On the left-hand-side, we use the integration formula

$$\int \frac{1}{x} dx = \ln x + \text{Constant}.$$

On the right-hand-side, we use the integration formula

$$\int dx = x + \text{Constant}.$$

After integrating both sides, we have

$$\ln [v_C(t) - 12] = -\frac{1}{RC} t + \text{Constant}.$$

or

$$v_C(t) - 12 = A e^{-\frac{1}{RC}t}, \quad \text{for } t \geq 0,$$

where A is a constant determined by the initial condition. The above expression can be rewritten as

$$v_C(t) = 12 + A e^{-\frac{1}{RC}t}, \quad \text{for } t \geq 0.$$

Since $v_C(0) = 0$, $A = -12$. Finally, we have

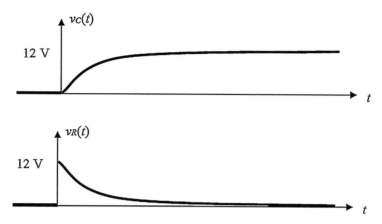

Fig. 17.2 At $t = 0$, the voltage across the resistor is discontinuous, while the voltage across the capacitor is continuous

$$v_C(t) = 12 - 12e^{-\frac{1}{RC}t}, \quad \text{for} \quad t \geq 0.$$

Hence,

$$v_R(t) = 12 - v_C(t) = 12e^{-\frac{1}{RC}t}, \quad \text{for } t > 0,$$

and

$$v_R(t) = 0, \quad \text{for} \quad t < 0.$$

Unlike a capacitor, the voltage across the resistor has a discontinuous jump at $t = 0$ from 0 V to 12 V. On the other hand, this voltage jump never happens for a capacitor because the energy stored in the capacitor takes time to charge or discharge (see Fig. 17.2).

Solving a differential equation is not an easy task in general. We will avoid differential equations as much as possible. This is the main motivation that we will use the phasor notation for sinusoidal steady-state analysis and use the Laplace transform for non-constant and non-steady-state cases as we will discover later.

For a first-order differential equation concerning only the DC power and switching actions, we can readily write down the solution for the capacitor voltage or inductor current if we know the initial value, final value, and the time constant.

Notes

Even though inductors and capacitors do not follow Ohm's law, KCL and KVL are still valid and can still be used to set up equations. The equations now are differential equations.

We can solve some simple differential equations. The solution of differential equations is a function of time (instead of a number).

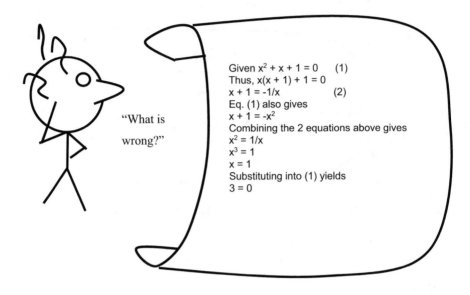

"What is

wrong?"

Given $x^2 + x + 1 = 0$ (1)
Thus, $x(x + 1) + 1 = 0$
$x + 1 = -1/x$ (2)
Eq. (1) also gives
$x + 1 = -x^2$
Combining the 2 equations above gives
$x^2 = 1/x$
$x^3 = 1$
$x = 1$
Substituting into (1) yields
$3 = 0$

Exercise Problems

Problem 17.1 Set up a node equation for the circuit in Fig. P17.1. Then express the equation in terms of i_L.

Fig. P17.1

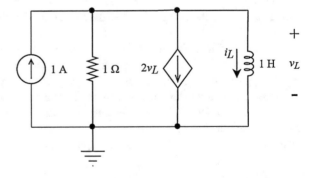

Problem 17.2 Set up a differential equation for $i_1 + i_2$.

Fig. P17.2

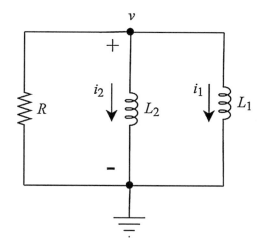

Problem 17.3 Set up a differential equation for $v_1 - v_2$.

Fig. P17.3

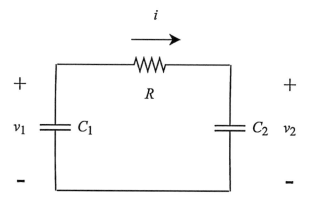

Solutions to Exercise problems are given in Book Appendix.

First-Order Circuits

18

Both RL and RC circuits are **first-order circuits** because their voltage and current can be related by a first-order differential equation. In Fig. 15.2, there is one inductor in the circuit; it is clearly a first-order circuit.

In Fig. 16.2, there are two capacitors in the circuit, which could be a second-order circuit. However, when the circuit is in action after the switch is open, those capacitors act independently. Therefore, the circuit in Fig. 16.2 can be decomposed into two first-order circuits after the switch is open.

For all first-order circuits, we can use the general exponential charging/discharging expression (see also Fig. 18.1)

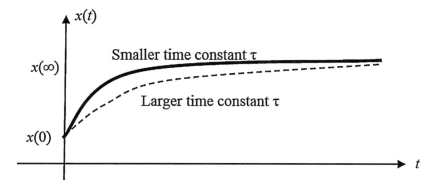

Fig. 18.1 For a first-order system, the response is an exponential function depending on the initial value, the final value, and the time constant. A smaller time constant gives a faster respond

$$x(t) = x(\infty) + [x(0) - x(\infty)]e^{-\frac{t}{\tau}}, \quad \text{for} \quad t \geq 0,$$

to find the time response by using three important values:

the **initial value** $x(0)$,
the **final value** $x(\infty)$, and
the **time constant** τ.

If solving a differential equation is challenging for you, you can substitute the above exponential function into the differential equation to see if it is really the solution. You can convince yourself whether the solution is correct by verification. Later you will learn the Laplace transform, which will convert a differential equation into an algebraic equation. An algebraic equation is easier to solve than a differential equation.

Notes

With simple switch actions, the first-order RC or RL circuits have an exponential response for the capacitor voltage or the inductor current. This exponential function is uniquely determined by the initial condition, the final condition, and the time constant.

$$x(t) = x(\infty) + [x(0) - x(\infty)]e^{-\frac{t}{\tau}}, \quad \text{for} \quad t \geq 0.$$

Exercise Problems

Problem 18.1 The input of an RC circuit is a periodic square pulse sequence. The period is $2T$. The time constant of the RC circuit is τ. The output signal is the capacitor voltage v_C. Find the output signal's maximum value v_{max} and the minimum value v_{min}.

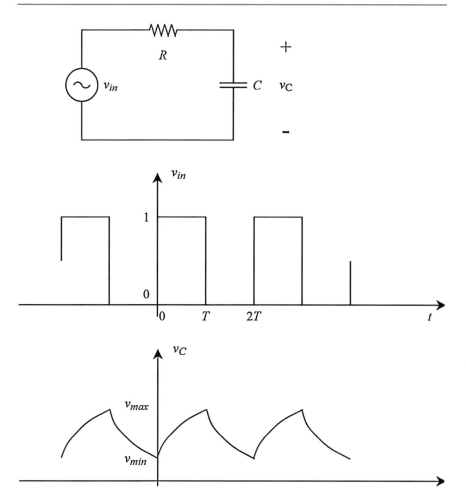

Fig. P18.1

Problem 18.2 A student tries to solve a problem in his own way, and he does not get the correct answer. Please help him to find the error. In the problem, the switch has been closed for a long time. The switch opens at $t = 0$. Find the capacitor's voltage i_C as function of time.

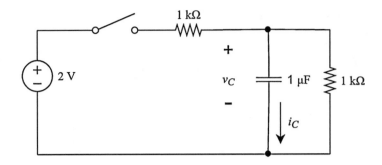

Fig. P18.2

The student's solution:
After the switch has been closed for a long time, the capacitor acts like an open circuit. Therefore,

$$i_C(0) = 0.$$

At $t = 0$, the switch opens. Now the circuit does not have any source, and the capacitor will eventually discharge to 0. Thus,

$$i_C(\infty) = 0.$$

Recall the general solution

$$i(t) = i(\infty) + [i(0) - i(\infty)]e^{-t/\tau}, \quad \text{for} \quad t \geq 0.$$

The student's solution is

$$i_C(t) = 0, \quad \text{for} \quad t \geq 0.$$

Solutions to Exercise problems are given in Book Appendix.

Sinusoidal Steady-State (Phasor) 19

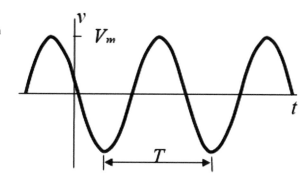

Fig. 19.1 A sine wave with $v = V_m \cos(\omega t + \varphi)$ with radian frequency $\omega = 2\pi f$ or frequency $f = \frac{1}{T}$

If the input of a linear circuit is a sine wave (see Fig. 19.1), the voltages and currents everywhere in the circuit are sine waves with the same radian frequency ω, but they may have different amplitudes V_m and phase angles φ.

Let us assume that the input source is a cosine function. We construct a complex function, $e^{j\omega t}$, by adding an imaginary part of a sine function. It happens that, in mathematics, this complex function is so special in a linear differential equation. It is an "**eigenfunction**." This eigenfunction has a property that the derivative of it is a constant $(j\omega)$ multiplication of it, and the integral of it is another constant $\left(-\frac{j}{\omega}\right)$ multiplication of it.

Therefore, derivative and integration can be performed by doing constant multiplication. No differential equations are needed. Since the eigenfunction $e^{j\omega t}$ appears in every expression, and do not need to carry it along with us when we work on equations. At the end, we convert the phasor notation back to the real world by putting the ωt thing back in and discarding the imaginary part.

"**Phasor**", $V_m \angle \varphi$, is a shorthand notation for $v = V_m \cos(\omega t + \varphi)$ with ωt omitted.

G. L. Zeng, M. Zeng, *Electric Circuits*,
https://doi.org/10.1007/978-3-030-60515-5_19

When the input is a sine wave, the circuit is an **AC (alternating current)** circuit. You do not treat an inductor as a short circuit and a capacitor as open circuit anymore. They are treated as "resistors" and characterized by a complex value known as the **impedance**. A common symbol for the impedance is Z.

By using impedance and phasor notation, Ohm's law works for capacitors and inductors as well as the resistors.

For a resistor:

$$v = R\,i.$$

Let

$$i = I_m \cos(\omega t),$$

and then

$$v = RI_m \cos(\omega t).$$

Phasor notation:

$$V = Z\,I,$$

with $Z = R$, $I = I_m \angle\, 0^\circ$, and $V = V_m \angle\, 0^\circ$.

For an inductor:

$$v = L\frac{di}{dt}.$$

Let

$$i = I_m \cos(\omega t).$$

Then

$$v = -\omega LI_m \sin(\omega t) = -\omega LI_m \cos(\omega t - 90).$$

Phasor notation:

$$V = (j\omega L)I = ZI,$$

with

$$Z = j\omega L = \omega L \angle 90^\circ.$$

$$\because e^{\frac{j\pi}{2}} = \cos\frac{\pi}{2} + j\sin\frac{\pi}{2} = 0 + j = j,$$

$$\therefore j = 1\angle 90^\circ$$

in the polar system.

For a capacitor:

$$i = C\frac{dv}{dt}.$$

Let

$$v = V_m \cos(\omega t),$$

and then

$$i = -\omega C V_m \sin(\omega t) = -\omega C V_m \cos(\omega t - 90).$$

Phasor notation:

$$I = (j\omega C)V = \frac{V}{Z}$$

with

$$Z = \frac{1}{j\omega C} = \frac{1}{\omega C}\angle -90^\circ.$$

$$\because e^{-\frac{j\pi}{2}} = \cos\frac{-\pi}{2} + j\sin\frac{-\pi}{2} = 0 - j = \frac{1}{j},$$

$$\therefore \frac{1}{j} = 1\angle -90^\circ$$

in the polar system.

What is a phasor anyway? A phasor $V_m \angle \varphi$ can represent $v = V_m \cos(\omega t + \varphi)$, but they are not the same thing. A phasor $V_m \angle \varphi$ is a complex number with a real part and an imaginary part, while $v = V_m \cos(\omega t + \varphi)$ is real.

Let us do some investigation to see how the phasor $V_m \angle \varphi$ is formed. Let us first form a complex number that uses $V_m \cos(\omega t + \varphi)$ as its real part and artificially adds an imaginary part $V_m \sin(\omega t + \varphi)$. This complex number can be expressed with an exponential function:

$$V_m \cos(\omega t + \varphi) + jV_m \sin(\omega t + \varphi) = V_m e^{j(\omega t + \varphi)} = V_m e^{j\omega t} e^{j\varphi}.$$

We then throw away $e^{j\omega t}$ from the expression above, obtaining our phasor

$$V_m \angle \varphi = V_m e^{j\varphi},$$

which is a complex number in the polar form (instead of the Cartesian form). By the way,

$$j = \sqrt{-1}.$$

If you want to multiply two complex numbers, it is easier to use the polar form. If you want to add two complex numbers, it is easier to use the Cartesian form. These two forms can be converted to each other by using **Euler's formula**:

$$e^{jx} = \cos x + j \sin x.$$

"This part is IMAGINARY; it is not really there, so that we can make the whole thing complex."

When you use impedance Z and phasor notation for voltages and currents, everything we discussed before works, including Ohm's law, KVL, KCL, node-voltage method, mesh-current method, Thévenin equivalent, and Norton equivalent.

Example

We use the circuit in Fig. 19.2 as an example to show how to Ohm's law for sinusoidal steady-state analysis. Find $i(t)$ with

$$v_S(t) = 80 \cos{(2000t)} \ V.$$

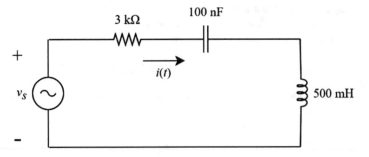

Fig. 19.2 An AC circuit with a capacitor, an inductor, and a resistor

Solution

Write the voltage source in phasor notation as

$$V_s = 80\angle 0°.$$

The three "resistors" are in series and the total impedance is

$$Z = R + j\omega L + \frac{1}{j\omega C},$$

$$= 3000 + j(2000)(0.5) + \frac{1}{j(2000)(100 \times 10^{-9})},$$

$$= 3000 - j4000 \; \Omega.$$

Ohm's law gives

$$I = \frac{V_s}{Z},$$

$$= \frac{80\angle 0°}{3000 - j4000},$$

$$= \frac{80\angle 0°}{\sqrt{3000^2 + 4000^2} \angle \tan^{-1} \frac{-4000}{3000}},$$

$$= 16\angle 53.13° \; \text{mA}.$$

Finally, change the phasor notation back to the normal time-domain notation

$$i(t) = 16\cos\left(2000t + 53.13°\right) \; \text{mA}.$$

Example

Given that

$$i_s(t) = 10\cos(50,000t) \; \text{A}$$

and

$$v_s(t) = 100\sin(50,000t) \; \text{V},$$

find the steady-state expression of $v(t)$ in the circuit of Fig. 19.3.

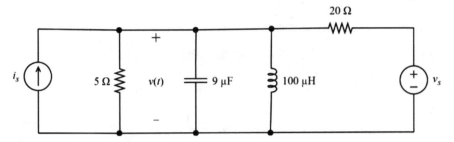

Fig. 19.3 An AC circuit with two sources

Solution

First, we rewrite the sources in phasor notation and convert the capacitance and inductance into impedance as shown in Fig. 19.4.

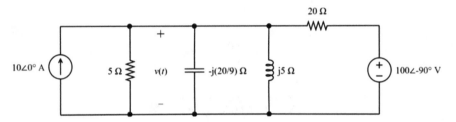

Fig. 19.4. The phasor notation is used for the circuit of Fig. 81

A node-voltage equation can be set up as

$$-10 + \frac{V}{5} + \frac{V}{-j\left(\frac{20}{9}\right)} + \frac{V}{j5} + \frac{V - 100\angle - 90^\circ}{20} = 0,$$

$$V = 10 - j30 = 31.62\angle - 71.57^\circ \text{ V}.$$

Therefore,

$$v(t) = 31.62\cos\left(50,000t - 71.57^\circ\right) \text{ V}.$$

Finally, we summarize this new concept of phasor.

When we do linear circuit analysis with sinusoidal inputs, we assume the steady state, which means that the power is on for a long time and the system is stabilized. In a linear system, the voltages and currents have the same frequency everywhere in the circuit, except for their amplitudes and phases. The phasor short-hand notation is a convenient way to represent the amplitude and phase, omitting the frequency which never changes.

The **phasor transform** is to add an imaginary part $V_m \sin(\omega t + \varphi)$ to the real function $V_m \cos(\omega t + \varphi)$, resulting in a complex function

$$V_m \cos(\omega t + \varphi) + jV_m \sin(\omega t + \varphi),$$

$$= V_m e^{j(\omega t + \varphi)},$$

$$= V_m e^{j\omega t} e^{j\varphi},$$

and to drop to frequency part $e^{j\omega t}$, obtaining a compact notation $V_m e^{j\varphi}$ or $V_m \angle \varphi$.

The **inverse phasor transform** is to write down the real function $V_m \cos(\omega t + \varphi)$ when you are given a complex number in the polar coordinate system $V_m \angle \varphi$. You can treat $V_m \angle \varphi$ as a normal complex number. You can multiply, divide, add, and subtract $V_m \angle \varphi$ with other complex numbers. When doing the inverse phasor transform, you must know the radian frequency ω because it is not given in the phasor notation.

The concept of impedance is an extension of resistance, to be used with the phasor notation. For an inductor, the impedance is

$$j\omega L = \omega L \angle 90^\circ.$$

For a capacitor, the impedance is

$$\frac{1}{j\omega C} = \frac{1}{\omega C} \angle -90^\circ.$$

Notes

In sinusoidal steady-state analysis, the frequency never changes. The phasor is a short-hand notation to represent the amplitude and phase of a sinusoidal function. A phasor is also a complex number expressed in the polar coordinate system.

$V_m \angle \varphi$ is a short-hand expression of $V_m \cos(\omega t + \varphi)$,

$$V_m \angle \varphi = V_m e^{j\varphi} = V_m \cos \varphi + jV_m \sin \varphi.$$

The inductor's impedance is $j\omega L$. The capacitor's impedance is $\frac{1}{j\omega C}$. When using impedance, Ohm's law works again.

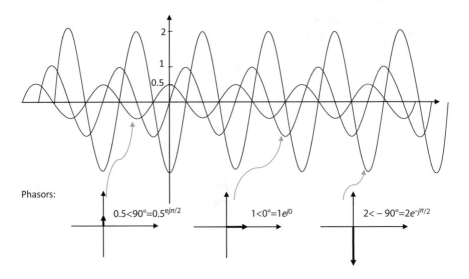

Phasors:

$0.5<90°=0.5e^{j\pi/2}$ $1<0°=1e^{j0}$ $2<-90°=2e^{-j\pi/2}$

Exercise Problems with Solutions

Problem 19.1 Express the following signals in the phasor form:

(a) $5 \cos (100t)$
(b) $5 \sin (100t)$
(c) $5 \cos (100t + 45°)$
(d) $5 \sin (100t + 45°)$
(e) $2 \cos (\omega t)$
(f) $2 \cos (\omega t) - 3 \cos (2\omega t)$
(g) 10
(h) $2t^2 \cos (\omega t)$

Problem 19.2 Express the transfer function in the phasor form. The input is v_{in} and the output is v_C.

Fig. P19.1

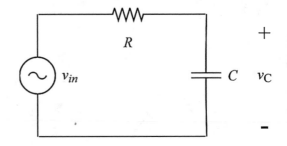

Solutions to Exercise problems are given in Book Appendix.

Function Generators and Oscilloscopes

20

A **function generator** and an **oscilloscope** are essential for us to do steady-state analysis of electric circuits. A function generator can produce periodic sine waves, square waves, triangle waves, sawtooth waves, and so on. It can perform other sophisticated tasks such as amplitude modulation. An oscilloscope can display the signals and do some measurements about them. This chapter serves as a tutorial for first-time users of these two pieces of equipment.

A function generator and an oscilloscope may look very similar. Figure 20.1 shows the front panel of a typical function generator. One way to tell a function generator from an oscillator is that a function generator has a section "Function" as shown in Fig. 20.1 that contains various function forms such as "Sine", "Square", and so on. You need to have a BNC (Bayonet Neill–Concelman)-alligator-clips cable with an impedance of 50 Ohms for the connection (see Fig. 20.2). Figure 20.3 shows how the alligator clips are connected to the circuit.

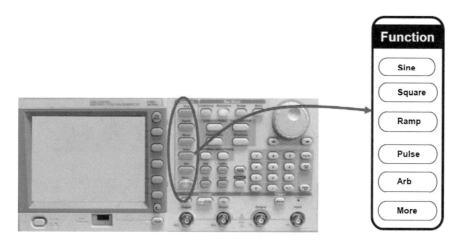

Fig. 20.1 A function generator

© The Author(s), under exclusive license to Springer Nature Switzerland AG 2021
G. L. Zeng, M. Zeng, *Electric Circuits*,
https://doi.org/10.1007/978-3-030-60515-5_20

Fig. 20.2 A BNC to alligator-clip cable for the signal generator output

Fig. 20.3 Connection of the
function generator to your
circuit

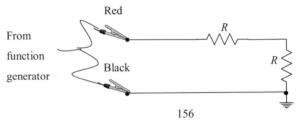

For example, if you want to generate a sine wave of 1 kHz. Do the following:

1. Make sure that the power cord is plugged in. Push the **on/off** switch to power on
   the equipment.

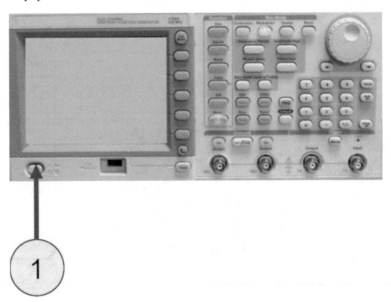

2. If you decide to use Channel 1 output of your function generator, connect a BNC
   cable to CH 1. The other end of the BNC cable can be connected to an oscilloscope to monitor the output of the function generator.

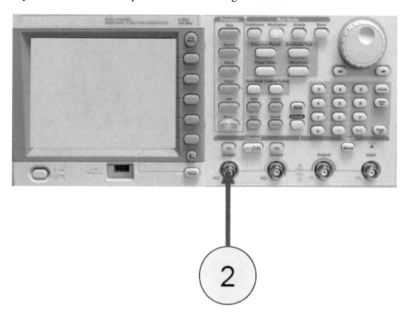

3. A sinewave can be selected by pushing the **Sine** button under "Function." Select
   the **Continuous** mode if it is not yet selected automatically when power is turned
   on.

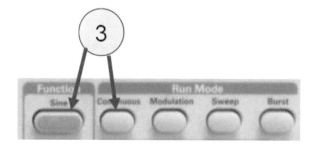

4. Push the front-panel CH1 Output **On** button to enable the output from channel 1 if your signal generator has more than one outputs. Some function generators only have one channel for output.

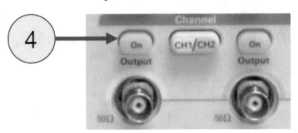

5. Use the oscilloscope auto-scaling function to display the sine waveform on the screen. An oscilloscope looks like a function generator. One way to tell an oscilloscope is that an oscilloscope has vertical and horizontal controls, as shown in Fig. 20.5.

Function Generator   Oscilloscope

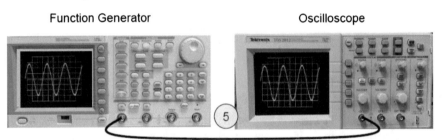

6. By pushing the **Frequency/Period** button you can get ready to set the frequency of the signal that you would like to generate.

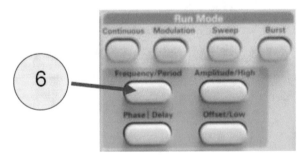

7. The **Frequency/Period** button toggles between Frequency setting and Period setting. You know the Frequency setting is selected if you see a small triangle on the screen pointing to **Freq**.

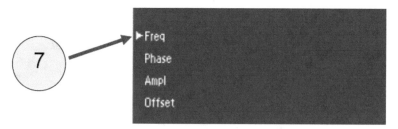

8. To set or change the frequency value, use the keypad to type in the numerical value and use the bezel buttons to specify the units.

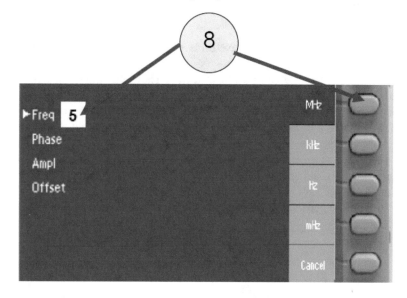

For example, you enter a value "5" from the keypad for the frequency value. At this time, the bezel menus will automatically change to Units. You can select the unit of the frequency that you desired, say, MHz.

You can change the Phase, Amplitude, and Offset values in the similar way.

9. Alternatively, you can change the values using the general-purpose turn-knob. By pushing the arrow keys and then turning the knob, you can change a specific digit.

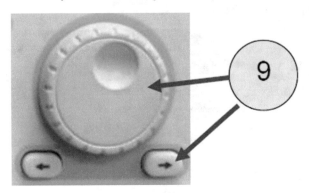

You can change to different waveforms by pushing different buttons on the control panel.

The function generator can be modeled as a Thévenin equivalent circuit (see Fig. 20.4), with $R_{\mathrm{Th}} = 50\,\Omega$. The settings of the function generator assume that your circuit matches the impedance $R_{\mathrm{Th}} = 50\,\Omega$. If you set the output amplitude of the signal generator as 5 V, the signal generator will produce a sine wave of 10 V. If your circuit's input impedance happens to be 50 Ω, you will get a 5 V sine wave. You are getting twice the voltage displayed on the function generator at the output terminal when the there is nothing connected to it. The voltage amplitude that you set on the function generator may not be what your circuit actually gets, depending on your circuit's input impedance. You can use an oscilloscope to measure the actual amplitude.

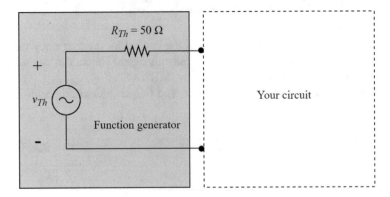

Fig. 20.4 The function generator has an output impedance of 50 Ω

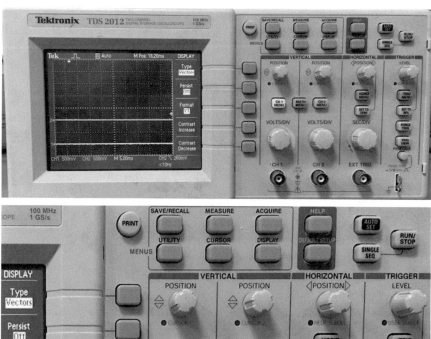

Fig. 20.5   An oscilloscope

The front panel of a typical oscilloscope is shown in Fig. 20.5. Special probes are required to use the oscilloscope (see Fig. 20.6). The oscilloscope has two input channels. We will use one channel to monitor the input signal, and the other channel to monitor the voltage at the node connecting two resistors (see Fig. 20.7 for connections).

You have many display options: Channel 1 only, Channel 2 only, both channels, and so on. **Coupling** refers to the method used to connect an electrical signal from your test circuit to the oscilloscope. DC coupling shows all components of an input signal. AC coupling, on the other hand, blocks the DC component of the signal.

The horizontal axis is the time axis on the display. The "Time/Div" knob controls the zoom scale on the time axis. The vertical axis is the voltage. The "Volts/Div" knobs control the zoom scale on the voltage axis.

Fig. 20.6 An oscilloscope
probe

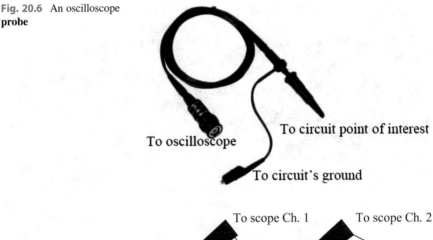

Fig. 20.7 Both the signal generator and the oscilloscope are connected to the circuit

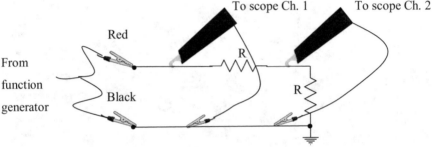

The scope probes are passive, and they do not contain any amplifiers. The $10\times$ setting of the probe attenuates the signal 10:1.

An oscilloscope's **trigger** function synchronizes the horizontal sweep at the correct point of the signal, essential for clear signal characterization (see Fig. 20.8). Trigger controls allow you to stabilize repetitive waveforms and capture single-shot waveforms.

The trigger makes repetitive waveforms appear static on the oscilloscope display by repeatedly displaying the same portion of the input signal. Imagine the jumble on the screen that would result if each sweep started at a different place on the signal, as illustrated in Fig. 20.9. The scope also has an auto mode, which can significantly reduce your frustration trying to get a stable display. An oscilloscope has many other useful functions. For example, you can move the cursor around to measure the time delay between two different signals and then you can convert the time delay to phase delay for your steady-state circuit analysis.

If this is your first time to use an oscilloscope, you can follow the procedure below to display a signal on the scope.

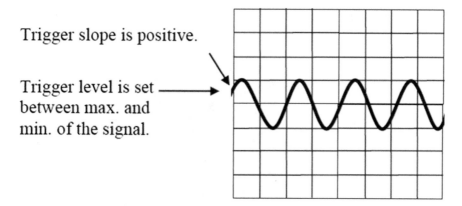

Trigger slope is positive.

Trigger level is set — between max. and min. of the signal.

Fig. 20.8 Trigger level must be properly chosen within the signal's voltage range

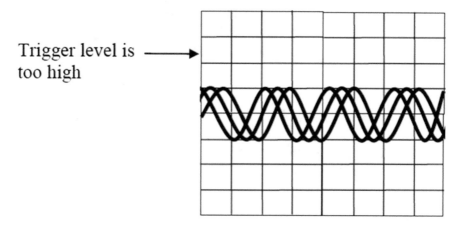

Trigger level is — too high

Fig. 20.9 An un-triggered display because the trigger level is selected too high

1. Connect the scope-probe to the input signal source, which can be a node in the circuit you are diagnosing.

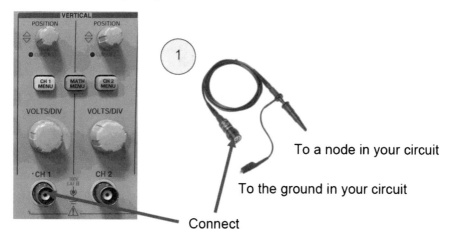

To a node in your circuit

To the ground in your circuit

Connect

2. Select the input channel (say, CH 1) that you connected in Step 1 by pushing the corresponding button.

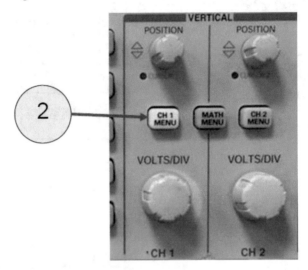

3. Press Auto-Set.

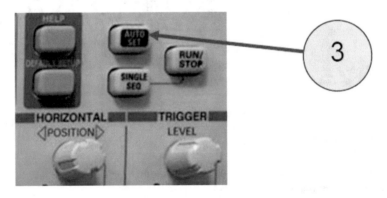

4. Adjust the vertical position and scale using the front-panel knobs. The vertical position knob moves your displayed signal up and down. The vertical Volts/Div knob scales the amplitude of your displayed signal.

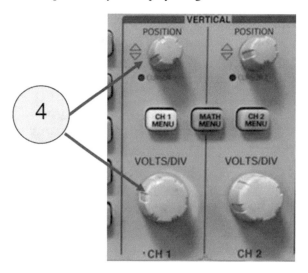

5. Adjust the horizontal position and scale using the front-panel knobs. The horizontal Volts/Div knob stretches or shrinks horizontally the signal displayed. The horizontal position knob determines the starting position of the signal displayed relative to trigger time.

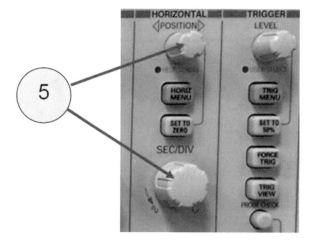

Notes

A function generator can generate periodic signals with a specified frequency and wave form. They are usually used as the sinusoidal power source for steady-state studies.

An oscilloscope can display the wave forms picked up from a circuit; it can be treated as a fancy graphic voltmeter. A proper triggering setup is required in order to have a stable display of a periodic wave form.

Exercise Problems

Problem 20.1 In Fig. 20.1, an oscilloscope is directly connected to a signal generator to verify the signal generated. We set the peak-to-peak voltage of a sinewave to be 10 V. However, the oscilloscope shows a 20 V peak-to-peak. Is there anything wrong?

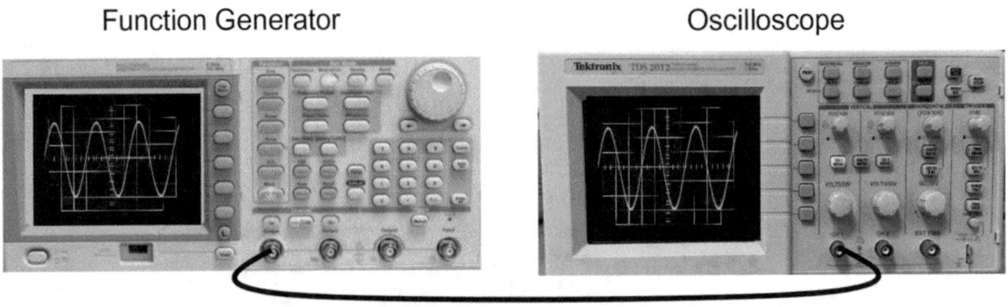

Function Generator   Oscilloscope

Fig. P20.1

Problem 20.2 How to use an oscilloscope to estimate the time constant of a first-order circuit?

Solutions to Exercise problems are given in Book Appendix.

Mutual Inductance and Transformers

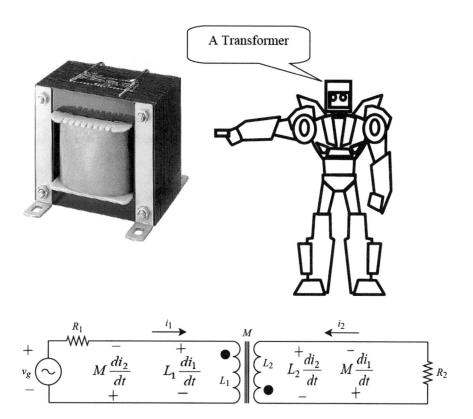

Fig. 21.1 In a transformer, two separate coils are magnetically coupled

When two separate coils are magnetically linked together, emf (voltage) can be generated by **mutual inductance**, M (see Fig. 21.1). The current in coil #1, i_1, can

© The Author(s), under exclusive license to Springer Nature Switzerland AG 2021   149
G. L. Zeng, M. Zeng, *Electric Circuits*,
https://doi.org/10.1007/978-3-030-60515-5_21

induce voltage in coli #2. Similarly, the current in coil #2, i_2, can induce voltage in coli #1. They are related by

$$v_2 = M \frac{di_1}{dt},$$

and

$$v_1 = M \frac{di_2}{dt}.$$

The induced voltage is created by the derivative of the current in the other coil. For DC current, the derivative is zero, and it will not induce any voltage. Therefore, a transformer does not work for DC.

Dot convention for mutually coupled coils:

- When the current enters the dotted terminal in one coil, the polarity of induced voltage in the other coil is positive at its dotted terminal.
- When the current leaves the dotted terminal in one coil, the polarity of induced voltage in the other coil is negative at its dotted terminal.

Using the mesh-current method, we have two differential equations for the circuit in Fig. 21.1:

$$-v_g + i_1 R_1 + L_1 \frac{di_1}{dt} - M \frac{di_2}{dt} = 0,$$

$$i_2 R_2 + L_2 \frac{di_2}{dt} - M \frac{di_1}{dt} = 0.$$

We can use the phasor notation to transform the above two differential equations into two algebraic equations:

$$-V_g + I_1 R_1 + j\omega L_1 I_1 - j\omega M I_2 = 0,$$

$$I_2 R_2 + j\omega L_2 I_2 - j\omega M I_1 = 0.$$

In the matrix form, the system of equations becomes

$$\begin{bmatrix} R_1 + j\omega L_1 & -j\omega M \\ -j\omega M & R_2 + j\omega L_2 \end{bmatrix} \begin{bmatrix} I_1 \\ I_2 \end{bmatrix} = \begin{bmatrix} V_g \\ 0 \end{bmatrix}.$$

Thus,

$$\begin{bmatrix} I_1 \\ I_2 \end{bmatrix} = \begin{bmatrix} R_1 + j\omega L_1 & -j\omega M \\ -j\omega M & R_2 + j\omega L_2 \end{bmatrix}^{-1} \begin{bmatrix} V_g \\ 0 \end{bmatrix}.$$

The solutions I_1 and I_2 are then transformed back to the time domain to get i_1 and i_2. Finally, the voltage across the resistor R_2 can be obtained via Ohm's law.

In everyday life, an ideal transformer is a good approximation for a practical transformer. A transformer is ideal if the coefficient of coupling $k = 1$, where $k = \frac{M}{\sqrt{L_1 L_2}}$ and $L_1 = L_2 \to \infty$. In this ideal situation, the primary and secondary voltages and currents are determined by the turn ratio, $N_1{:}N_2$, of the transformer.

In an **ideal transformer** (see Fig. 21.2), the magnitude of the volts per turn is the same for each coil, or

$$\left| \frac{V_1}{N_1} \right| = \left| \frac{V_2}{N_2} \right|.$$

The magnitude of the ampere-turns is the same for each coil, or

$$|I_1 N_1| = |I_2 N_2|.$$

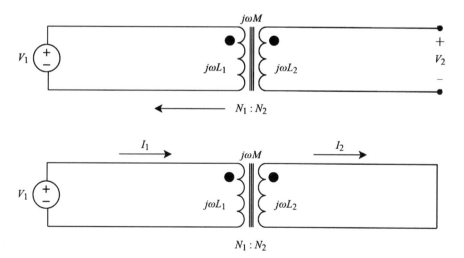

Fig. 21.2 Ideal transformer (phasor notation)

Even though the most popular application of transformers is to change AC voltages, another important application is **impedance matching**, when the maximum power transfer is desired.

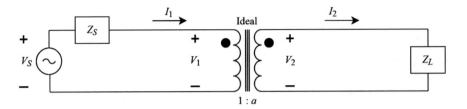

Fig. 21.3 A transform is used for impedance matching

Referring to Fig. 21.3 and considering an ideal transformer, we have

$$V_2 = aV_1,$$

$$I_1 = aI_2.$$

The **reflected impedance** of Z_L virtually on the primary side is

$$Z_{\text{reflected}} = \frac{V_1}{I_1} = \frac{1}{a^2}\frac{V_2}{I_2} = \frac{1}{a^2}Z_L.$$

For example, if we have an audio amplifier with internal impedance Z_s of 800 Ω and a speaker Z_L of 8 Ω, the speaker does not get much power if it is directly connected to the audio amplifier output. The use of a 10:1 step-down transformer (i.e., $a = 0.1$) in between can greatly improve the power output. The maximum power condition is

$$Z_S = Z^*_{\text{reflected}}.$$

Notes

A basic transformer consists of two coils, sharing the same magnetic field. The mutual inductance helps to induce an emf (voltage) in an adjacent coil. The induced emf is proportional to the rate of change in the current. Therefore, a transformer does not work for DC currents.

The turn ratio of a transformer determines the voltage ratio and current ratio.

A transformer can also be used for impedance matching.

Exercise Problems

Problem 21.1 An ideal transformer has 1000 turns in its primary coil and 100 turns in its secondary coil. Determine whether the following statements are true.

(a) This is a 10:1 transformer.
(b) This is a 1:10 transformer.
(c) This is a 1:0.1 transformer.
(d) This is a 0.1:1 transformer.
(e) This is a step-up transformer.
(f) This is a step-down transformer.
(g) If the primary voltage is 10 V, the secondary voltage is 100 V.
(h) If the primary voltage is 10 V, the secondary voltage is 1 V.
(i) If the primary current is 10 A, the secondary current is 100 A.
(j) If the primary current is 10 A, the secondary current is 1 A.
(k) The transformer consumes power.
(l) The transformer only works for a DC source input.
(m) The transformer only works for an AC source input.
(n) The frequency on secondary side is ten times higher than the frequency on the primary side.
(o) The frequency on secondary side is ten times lower than the frequency on the primary side.
(p) If the secondary side has a load of 100 Ω, the reflected impedance on the primary side is 1000 Ω.
(q) If the secondary side has a load of 100 Ω, the reflected impedance on the primary side is 10,000 Ω.

Problem 21.2 This problem about the dot notation and convention in a transformer. Express the induced voltages for each case.

(a)

Fig. P21.1

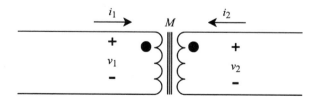

(b)

Fig. P21.2

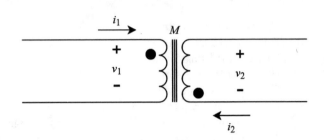

(c)

Fig. P21.3

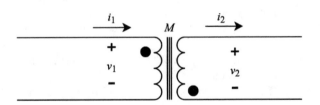

(d)

Fig. P21.4

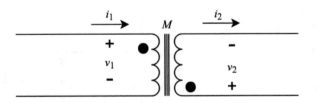

(e)

Fig. P21.5

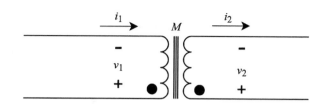

(f)

Fig. P21.6

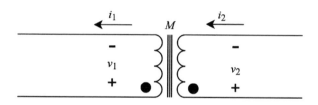

(g)

Fig. P21.7

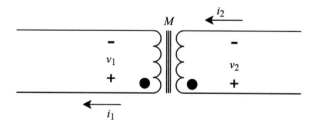

(h)

Fig. P21.8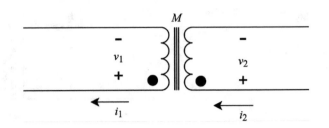

Solutions to Exercise problems are given in Book Appendix.

Fourier Series

22

We know how to do steady-state linear circuit analysis with sinusoidal inputs by using phasor notation and impedance. What shall we do if the voltage source produces a train of square waves or triangle waves, instead of sine waves? One way to solve this problem is to convert the periodic non-sinusoidal waves into sinusoidal waves. This procedure is called **Fourier series** expansion.

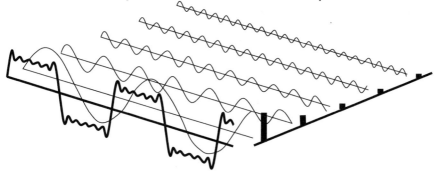

The theory of Fourier series expansion is that every **periodic function** can be expressed as a combination of multiple sine waves

$$f(t) = a_0 + \sum_{n=1}^{\infty} [a_n \cos(n\omega_0 t) + b_n \sin(n\omega_0 t)],$$

where $f(t) = f(t + T)$ is a periodic function with a period T,

$\omega_0 = \frac{2\pi}{T}$ is the fundamental frequency,
$2\omega_0, 3\omega_0, 4\omega_0, \cdots$ are the harmonic frequencies,
and a_n and b_n are the Fourier coefficients.

© The Author(s), under exclusive license to Springer Nature Switzerland AG 2021
G. L. Zeng, M. Zeng, *Electric Circuits*,
https://doi.org/10.1007/978-3-030-60515-5_22

We now use an example to illustrate how to use the Fourier expansion method to do steady-state circuit analysis with a non-sinusoidal input. Figure 22.1 gives a periodic function $f(t)$ with a period T. Assume that

$$f(t) = \sin(\omega_0 t) + \sin(3\omega_0 t).$$

Its fundamental wave is a sine function $sin(\omega_0 t)$ with the same period T as $f(t)$. Its third harmonic wave $sin(3\omega_0 t)$ has a period of $T/3$. This function $f(t)$ can be represented by the superposition of two sine waves, one with the fundament frequency and the other with the third harmonic wave.

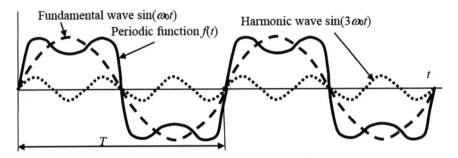

Fig. 22.1 A periodic function $f(t)$, its fundament wave, and its third harmonic wave

When doing circuit analysis, the two circuits in Fig. 22.2 are equivalent. You can either use the source $f(t)$ or use its expansion; you will get the same answer. Our goal here is to find the voltage across the capacitor in the circuit.

There are two voltage sources in Fig. 22.2 (right), and these two sources have different frequencies. We do not know how to handle two different frequencies simultaneously when doing steady-state analysis with the phasor notation.

Our approach here is to use the superposition principle and to consider one frequency at a time as shown in the circuits in Fig. 22.3 in the phasor notation. The output voltage in both circuits in Fig. 22.3 can be found by using the voltage divider.

$$V_1 = \left(1 \angle -90^\circ\right) \frac{\frac{1}{j\omega_0 C}}{R + j\omega_0 L + \frac{1}{j\omega_0 C}},$$

$$= -j \frac{\frac{1}{j\omega_0 C}}{R + j\omega_0 L + \frac{1}{j\omega_0 C}},$$

$$= j \frac{1}{j\omega_0 RC - \omega_0^2 LC + 1},$$

$$= \frac{1}{\sqrt{\left(1 - \omega_0^2 LC\right)^2 + \left(\omega_0 RC\right)^2}} \angle \left(-90^\circ - \tan^{-1} \frac{\omega_0 RC}{1 - \omega_0^2 LC}\right).$$

Similarly,

$$V_2 = \left(1 \angle - 90^\circ\right) \frac{\frac{1}{j3\omega_0 C}}{R + j3\omega_0 L + \frac{1}{j3\omega_0 C}},$$

$$= \frac{1}{\sqrt{\left(1 - 9\omega_0^2 LC\right)^2 + \left(3\omega_0 RC\right)^2}} \angle \left(-90^\circ - \tan^{-1} \frac{3\omega_0 RC}{1 - 9\omega_0^2 LC}\right).$$

After the **inverse phasor transform**, we have

$$v(t) = v_1(t) + v_2(t),$$

$$= \frac{1}{\sqrt{\left(1 - \omega_0^2 LC\right)^2 + \left(\omega_0 RC\right)^2}}$$

$$\times \cos\left(\omega_0 t + \angle\left(-90^\circ - \tan^{-1} \frac{\omega_0 RC}{1 - \omega_0^2 LC}\right)\right)$$

$$+ \frac{1}{\sqrt{\left(1 - 9\omega_0^2 LC\right)^2 + \left(3\omega_0 RC\right)^2}}$$

$$\times \cos\left(3\omega_0 t + \angle\left(-90^\circ - \tan^{-1} \frac{3\omega_0 RC}{1 - 9\omega_0^2 LC}\right)\right),$$

$$= \frac{1}{\sqrt{\left(1 - \omega_0^2 LC\right)^2 + \left(\omega_0 RC\right)^2}} \sin\left(\omega_0 t - \tan^{-1} \frac{\omega_0 RC}{1 - \omega_0^2 LC}\right)$$

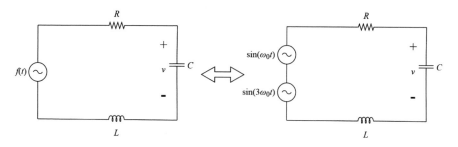

Fig. 22.2 Since $f(t) = \sin(\omega_0 t) + \sin(3\omega_0 t)$, these two circuits are equivalent

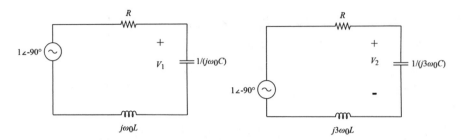

Fig. 22.3 The circuit of Fig. 22.2 (right) is converted into the phasor notation and is decomposed into two circuits with two different frequencies

$$+\frac{1}{\sqrt{\left(1 - 9\omega_0^2 LC\right)^2 + (3\omega_0 RC)^2}} \sin\left(3\omega_0 t - \tan^{-1}\frac{3\omega_0 RC}{1 - 9\omega_0^2 LC}\right).$$

Our next question is how to find the Fourier expansion when we are given a general periodic function $f(t)$. We need a set of formulas to find the Fourier coefficients a_n and b_n:

$$a_n = \int_T f(t) \cos(n\omega_0 t) dt, \quad n = 0, 1, 2, \ldots,$$

$$b_n = \int_T f(t) \sin(n\omega_0 t) dt, \quad n = 1, 2, \ldots,$$

where $\int_T$ means integration over one period T. That is, $\int_T = \int_{t_0}^{t_0+T}$ for any user preferred value t_0.

We now show you how these two formulas for a_n and b_n are derived by using the orthogonality method. **Orthogonality** means two vectors are perpendicular (i.e., $90°$) to each other. One simple example is the x–y coordinate system, in which the x-axis and y-axis are **orthogonal** to each other as seen in Fig. 22.4.

By inspection of Fig. 22.4, vector $(0, 3)$ is orthogonal to vector $(3, 0)$. Vector $(1, 1)$ is orthogonal to vector $(-2, 2)$. In mathematics, two vectors $\vec{v}_1 = (x_1, y_1)$ and $\vec{v}_2 = (x_2, y_2)$ are orthogonal if their **inner product** (or, **dot product**) is zero, that is,

$$< \vec{v}_1, \vec{v}_2 >= x_1 x_2 + y_1 y_2 = 0.$$

We can readily verify that

$$< (0, 3), (3, 0) >= 0 \times 3 + 3 \times 0 = 0$$

and

$$< (1, 1), (-2, 2) >= 1 \times (-2) + 1 \times 2 = 0.$$

Fig. 22.4 The x–y coordinate system

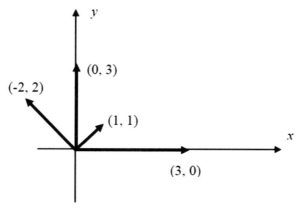

The inner product can be extended for two real one-dimensional (1D) functions f and g by using the integration-defined inner product as

$$< f, g >= \int f(t)g(t)\mathrm{d}t.$$

Two functions f and g are orthogonal if their inner product is 0. If the two functions f and g are periodic with a period T, the integration interval is from t_0 to $t_0 + T$, for an arbitrary value t_0.

It is not difficult to verify that $cos(n\omega_0 t)$ and $cos(m\omega_0 t)$ are orthogonal, and $sin(n\omega_0 t)$ and $sin(m\omega_0 t)$ are orthogonal, if $n \neq m$. We also have $cos(n\omega_0 t)$ and $sin(m\omega_0 t)$ are orthogonal, for all n and m. In fact,

$$\int_{t_0}^{t_0+T} \mathrm{d}t = T.$$

$$\int_{t_0}^{t_0+T} \sin(n\omega_0 t)\mathrm{d}t = 0, \quad \text{for all integers } n.$$

$$\int_{t_0}^{t_0+T} \cos(n\omega_0 t)\mathrm{d}t = 0, \quad \text{for all non} - \text{zero integers } n.$$

$$\int_{t_0}^{t_0+T} \cos(n\omega_0 t)\sin(m\omega_0 t)\mathrm{d}t = 0, \quad \text{for all } m \text{ and } n.$$

$$\int_{t_0}^{t_0+T} \cos(n\omega_0 t)\cos(m\omega_0 t)\mathrm{d}t = \begin{cases} 0, & \text{for } m \neq n \\ \dfrac{T}{2}, & \text{for } m = n \end{cases}$$

$$\int_{t_0}^{t_0+T} \sin (n\omega_0 t) \sin (m\omega_0 t)\mathrm{d}t = \begin{cases} 0, & \text{for} \quad m \neq n \\ \dfrac{T}{2}, & \text{for} \quad m = n. \end{cases}$$

To find a_0, we integrate both sides of

$$f(t) = a_0 + \sum_{n=1}^{\infty} [a_n \cos (n\omega_0 t) + b_n \sin (n\omega_0 t)]$$

over T:

$$\int_{t_0}^{t_0+T} f(t)\mathrm{d}t = \int_{t_0}^{t_0+T} a_0 \mathrm{d}t,$$

$$+ \sum_{n=1}^{\infty} \int_{t_0}^{t_0+T} [a_n \cos (n\omega_0 t) + b_n \sin (n\omega_0 t)]\mathrm{d}t,$$

$$\int_{t_0}^{t_0+T} f(t)\mathrm{d}t = Ta_0 + 0.$$

Thus,

$$a_0 = \frac{1}{T} \int_{t_0}^{t_0+T} f(t)\mathrm{d}t.$$

To find a_n, we multiply both sides of

$$f(t) = a_0 + \sum_{m=1}^{\infty} [a_n \cos (m\omega_0 t) + b_n \sin (m\omega_0 t)]$$

with $\cos(n\omega_0 t)$ and then integrate both sides over T:

$$\int_{t_0}^{t_0+T} f(t) \cos (n\omega_0 t)\mathrm{d}t,$$

$$= \int_{t_0}^{t_0+T} a_0 \cos (n\omega_0 t)\mathrm{d}t$$

$$+ \sum_{m=1}^{\infty} \int_{t_0}^{t_0+T} [a_n \cos (m\omega_0 t) \cos (n\omega_0 t) + b_n \sin (m\omega_0 t) \cos (n\omega_0 t)]\mathrm{d}t,$$

$$\int_{t_0}^{t_0+T} f(t) \cos{(n\omega_0 t)}dt = 0 + \frac{T}{2}a_n + 0.$$

Thus,

$$a_n = \frac{2}{T}\int_{t_0}^{t_0+T} f(t) \cos{(n\omega_0 t)}dt.$$

To find b_n, we multiply both sides of

$$f(t) = a_0 + \sum_{m=1}^{\infty} [a_n \cos{(m\omega_0 t)} + b_n \sin{(m\omega_0 t)}]$$

with $\sin(n\omega_0 t)$ and then integrate both sides over T:

$$\int_{t_0}^{t_0+T} f(t) \sin{(n\omega_0 t)}dt$$

$$= \int_{t_0}^{t_0+T} a_0 \sin{(n\omega_0 t)}dt$$

$$+ \sum_{m=1}^{\infty} \int_{t_0}^{t_0+T} [a_n \cos{(m\omega_0 t)} \sin{(n\omega_0 t)} + b_n \sin{(m\omega_0 t)} \sin{(n\omega_0 t)}]dt,$$

$$\int_{t_0}^{t_0+T} f(t) \sin{(n\omega_0 t)}dt = 0 + 0 + \frac{T}{2}b_n.$$

Thus,

$$b_n = \frac{2}{T}\int_{t_0}^{t_0+T} f(t) \sin{(n\omega_0 t)}dt.$$

Before we proceed to find a Fourier expansion of a given periodic function, we have some short cuts.

1. If $f(t)$ has no **DC offset**, $a_0 = 0$.
2. If $f(t)$ is an **even function**, then all $b_n = 0$ because $f(t)$ can only be built with even functions $\cos(n\omega_0 t)$.
3. If $f(t)$ is an **odd function**, then all $a_n = 0$ because $f(t)$ can only be built with odd functions $\sin(n\omega_0 t)$.
4. If $f(t)$ has **half-wave symmetry**, that is, $f(t) = -f\left(t - \frac{T}{2}\right)$ [Delay $T/2$, becomes negative], then even coefficients are zero. That is,

$$a_n = b_n = 0, \text{ for } n \text{ even.}$$

An example of half-wave symmetry is shown in Fig. 22.5.

Fig. 22.5 A function with
half-wave symmetry

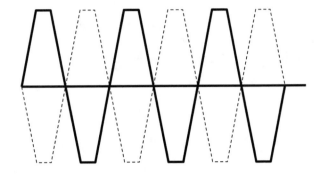

5. As a special case of (3), if the half-wave itself is symmetric as indicated in
   Fig. 22.6, the Fourier coefficient integral can be evaluated only over $T/4$:

$$a_n = \frac{8}{T} \int_0^{\frac{T}{4}} f(t) \cos (n\omega_0 t)\mathrm{d}t,$$

$$b_n = \frac{8}{T} \int_0^{\frac{T}{4}} f(t) \sin (n\omega_0 t)\mathrm{d}t.$$

Fig. 22.6 A function with
half-wave symmetry

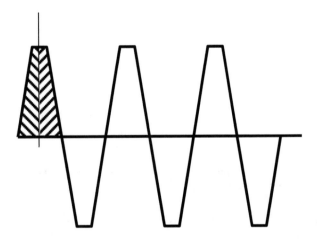

Example

Find the Fourier expansion for the periodic function defined in Fig. 22.7.

Fig. 22.7 A periodic triangle
function $f(t)$

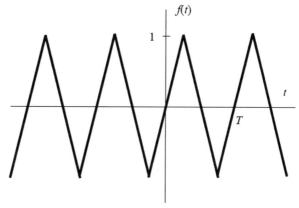

Solution
This function satisfies conditions (1), (3), (4), and (5). All we need to find is b_n for
n being odd. We can write down the equation for the straight-line segment of $f(t)$ on
$[0, T/4]$ as

$$f(t) = \frac{4}{T}t, t \in \left[0, \frac{T}{4}\right].$$

Then

$$b_n = \frac{8}{T} \int_0^{\frac{T}{4}} f(t) \sin (n\omega_0 t)\,dt,$$

$$= \frac{8}{T} \int_0^{\frac{T}{4}} \frac{4}{T}t \sin (n\omega_0 t)\,dt,$$

$$= \frac{32}{T^2} \left(\frac{\sin (n\omega_0 t)}{n^2\omega_0^2} - \frac{t \cos (n\omega_0 t)}{n\omega_0} \right) \Bigg|_0^{\frac{T}{4}}, \text{ with } \omega_0 = \frac{2\pi}{T},$$

$$= \frac{8}{\pi^2 n^2} \sin \frac{n\pi}{2}, \text{ with } n \text{ being odd}.$$

The Fourier series for $f(t)$ is given as

$$f(t) = \frac{8}{\pi^2} \sum_{n=1,3,5,\dots}^{\infty} \frac{1}{n^2} \sin \frac{n\pi}{2} \sin (n\omega_0 t),$$

$$= \frac{8}{\pi^2} \left[\sin (\omega_0 t) - \frac{1}{9} \sin (3\omega_0 t) + \frac{1}{25} \sin (5\omega_0 t) - \frac{1}{49} \sin (7\omega_0 t) + \dots \right].$$

Last but not least, the Fourier expansion can also be used to find the power of the function by using the **Parseval's theorem**:

$$\frac{1}{T}\int_0^T [f(t)]^2 dt = a_0^2 + \frac{1}{2}\sum_{n=1}^{\infty}\left(a_n^2 + b_n^2\right).$$

The proof of the Parseval's theorem is rather straight forward. The starting point is the expansion

$$f(t) = a_0 + \sum_{n=1}^{\infty}\left[a_n \cos\left(n\omega_0 t\right) + b_n \sin\left(n\omega_0 t\right)\right].$$

We square both sides and then integrate both sides over T. Using the orthogonality, the integration results are

$$\int_T [f(t)]^2 dt = \int_T \left\{a_0 + \sum_{n=1}^{\infty}\left[a_n \cos\left(n\omega_0 t\right) + b_n \sin\left(n\omega_0 t\right)\right]\right\}^2 dt,$$

$$\int_T [f(t)]^2 dt = \int_T a_0^2 dt$$

$$+ \sum_{n=1}^{\infty}\left[a_n^2 \int_T \cos^2\left(n\omega_0 t\right) dt + b_n^2 \int_T \sin^2\left(n\omega_0 t\right) dt\right],$$

$$\int_T [f(t)]^2 dt = a_0^2 T + \frac{T}{2}\sum_{n=1}^{\infty}\left(a_n^2 + b_n^2\right).$$

Thus,

$$\frac{1}{T}\int_0^T [f(t)]^2 dt = a_0^2 + \frac{1}{2}\sum_{n=1}^{\infty}\left(a_n^2 + b_n^2\right).$$

Notes
Any periodic function can be represented as a combination of sine waves with different frequencies. The period of the function determines the fundamental frequency, which is the lowest frequency in the Fourier expansion. Other frequencies (called harmonics) are positive integer multiples of the fundamental frequency.

The harmonic functions are orthogonal to each other. Fourier expansion coefficient formulas are derived based on the orthogonality properties.

Exercise Problems

Problem 22.1 Match a Fourier series with a periodic function with $\omega_0 = 2\pi/T$.

$$f_1(t) = a_0 + \sum_{n=1}^{\infty} a_n \cos(n\omega_0 t),$$

$$f_2(t) = \sum_{n=1}^{\infty} a_n \cos(n\omega_0 t)$$

$$f_3(t) = a_0 + \sum_{n=1}^{\infty} b_n \sin(n\omega_0 t),$$

$$f_4(t) = \sum_{n=1}^{\infty} b_n \sin(n\omega_0 t),$$

$$f_5(t) = \sum_{\substack{n=1 \\ n=odd}}^{\infty} [a_n \cos(n\omega_0 t) + b_n \sin(n\omega_0 t)],$$

$$f_6(t) = \sum_{\substack{n=1 \\ n = odd}}^{\infty} a_n \cos(n\omega_0 t),$$

$$f_7(t) = \sum_{\substack{n=1 \\ n = odd}}^{\infty} b_n \sin(n\omega_0 t),$$

$$f_8(t) = a_0 + \sum_{n=1}^{\infty} [a_n \cos(n\omega_0 t) + b_n \sin(n\omega_0 t)],$$

$$f_9(t) = \sum_{n=1}^{\infty} [a_n \cos(n\omega_0 t) + b_n \sin(n\omega_0 t)].$$

(a)

Fig. P22.1

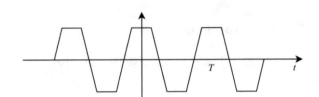

(b)

Fig. P22.2

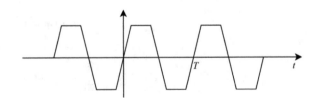

(c)

Fig. P22.3

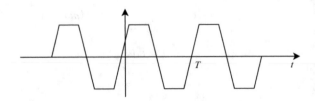

(d)

Fig. P22.4

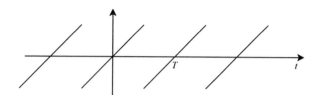

(e)

Fig. P22.5

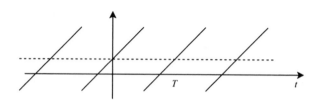

(f)

Fig. P22.6

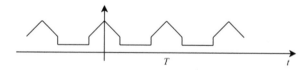

(g)

Fig. P22.7

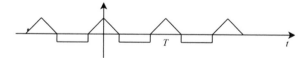

(h)

Fig. P22.8

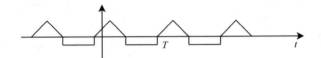

(i)

Fig. P22.9

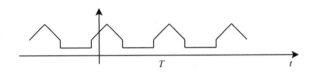

Problem 22.2 Show that the Fourier series has an equivalent exponential form

$$f(t) = \sum_{n=-\infty}^{\infty} c_n e^{jn\omega_0 t}.$$

Solutions to Exercise problems are given in Book Appendix.

Laplace Transform in Circuit Analysis

23

Why do we need this **Laplace transform**? The reason is that we do not want to solve differential equations when the circuit contains capacitors and/or inductors. Capacitors and inductors are troublemakers, making mathematical work harder than necessary. We already have some methods to deal with them. The Laplace transform method is just an alternative one.

For the first-order system, if the power sources are DC with some switch actions, the first-order differential equation is easy to solve, and the solution is essentially an exponential function that can be uniquely determined by three important values: the initial value, the final value, and the time constant.

If the sources are sine waves and the steady-state solutions are to be sought, the situation is fairly easy if the phasor notation is used. This phasor approach only works for sinusoidal sources and for steady-state analysis. What if the sources are periodical, but not exactly sinusoidal? This is where the Fourier series expansion comes in. Every periodic function can be represented by a Fourier series

$$f(t) = a_0 + \sum_{n=1}^{\infty} [a_n \cos (n\omega_0 t) + b_n \sin (n\omega_0 t)].$$

In fact, this equation does not hold at the points where $f(t)$ is not continuous. At the jumping points, the Fourier series converges to the average of the left and right limits of $f(t)$ at those points. In this tutorial, let us do not worry about those discontinuities. Once the Fourier series expansion is obtained, the phasor method can apply by using the superposition principle, considering one sine wave at a time.

What is the Laplace transform good for? It is good for the sources that are turned on at $t = 0$. Obviously, it will not work for any steady-state analysis. The Laplace transform converts a real function $f(t)$, $t \geq 0$ to a complex function $F(s)$, where s is a complex number. The derivative of $f(t)$ will correspond to $sF(s) - f(0^-)$, where $f(0^-)$ is the initial condition of the function $f(t)$ right before the source is turned on at $t = 0$. We have some important Laplace transform properties listed in Table 23.1.

© The Author(s), under exclusive license to Springer Nature Switzerland AG 2021
G. L. Zeng, M. Zeng, *Electric Circuits*,
https://doi.org/10.1007/978-3-030-60515-5_23

Table 23.1 Some properties of the laplace transform	$f(t), t \geq 0$	$F(s) =$ Laplace transform of $f(t)$
	$kf(t)$	$kF(s)$
	$f_1(t) + f_2(t)$	$F_1(s) + F_2(s)$
	$f'(t)$	$sF(s) - f(0^-)$
	$f''(t)$	$s^2F(s) - sf(0^-) - f'(0^-)$
	$f'''(t)$	$s^3F(s) - s^2f(0^-) - sf'(0^-) - f''(0^-)$
	$\int_0^t f(x)dx$	$\frac{1}{s}F(s)$

Recall that for the inductor, the current, and voltage are related by

$$v_L = L\frac{di_L}{dt}.$$

Its Laplace domain version is (by using Table 23.1 and assuming zero initial condition)

$$V_L = (sL)I_L.$$

This Laplace domain expression is in the form of Ohm's law with the impedance

$$Z_L = sL,$$

which is similar to the impedance $j\omega$ with the phasor notation. In fact, you can just replace $j\omega$ with s, you will get the Laplace transform version from the phasor method's impedance expression.

Let us do the same for the capacitor, for which the current and voltage are related by

$$i_C = C\frac{dv_C}{dt}.$$

Its Laplace domain version is (by using Table 23.1 and assuming zero initial condition)

$$I_C = (sC)V_C.$$

This Laplsace domain expression is in the form of Ohm's law with the impedance

$$Z_C = \frac{1}{sC},$$

which is similar to the impedance $\frac{1}{j\omega C}$ with the phasor notation. Therefore, by using the Laplace transform, we can feel freely to use Ohm's law to set up circuit equations. Before we can show you a circuit example, we need some **Laplace transform pairs**, some of which are listed in Table 23.2.

Table 23.2 Laplace transform pairs

$f(t), t \geq 0$	$F(s) =$ Laplace transform of $f(t)$
$\delta(t)23$	1
$u(t)$	$\frac{1}{s}$
$tu(t)$	$\frac{1}{s^2}$
$e^{-at}u(t)$	$\frac{1}{s+a}$
$\sin(\omega t)u(t)$	$\frac{\omega}{s^2+\omega^2}$
$\cos(\omega t)u(t)$	$\frac{s}{s^2+\omega^2}$
$te^{-at}u(t)$	$\frac{1}{(s+a)^2}$
$e^{-at}\sin(\omega t)u(t)$	$\frac{\omega}{(s+a)^2+\omega^2}$
$e^{-at}\cos(\omega t)u(t)$	$\frac{s+a}{(s+a)^2+\omega^2}$

In Table 23.2, $u(t)$ is the **unit step function**, defined as (see Fig. 23.1)

$$u(t) = \begin{cases} 0 & \text{for} \quad t < 0, \\ 0.5 & \text{for} \quad t = 0, \\ 1 & \text{for} \quad t > 0. \end{cases}$$

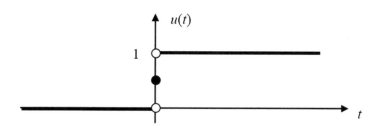

Fig. 23.1 The unit step function

In Table 23.2, $\delta(t)$ is called a **unit impulse function**, or **Dirac delta function**, or simply **delta function**. The delta function is not really a function in the regular sense. Anyway, we treat it as a function in this tutorial. It can be defined by the following two equations:

$$\delta(t) = \begin{cases} +\infty, & t = 0, \\ 0, & t \neq 0, \end{cases}$$

and

$$\int_{-\infty}^{\infty} \delta(t)dt = 1.$$

It is difficult to graphically show this function. Imagine that ε is a very small positive number, as we take the limit $\varepsilon \to 0$, the limit of Fig. 23.2 (left) tends to the delta function. Notice that the area underneath the function curve is always 1. An official symbol for the delta function is an arrow of length 1 as shown in Fig. 23.2 (right).

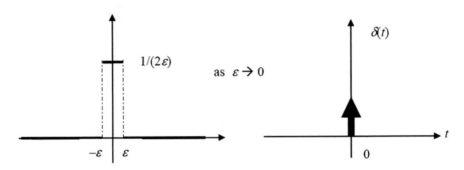

Fig. 23.2 The delta function $\delta(t)$

Example
Find the voltage $v_C(t)$ after $t = 0$.

Fig. 23.3 An RC circuit

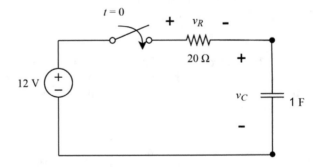

Solution
After $t = 0$, a voltage source of 12 V is applied. This is equivalent to a unit step function with amplitude 12 V. Thus, the Laplace domain version of the circuit in Fig. 23.3 is shown in Fig. 23.4.

Fig. 23.4 The Laplace
domain version of the circuit
in Fig. 23.3

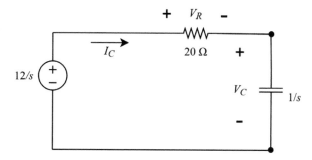

This is a voltage divider, and thus

$$V_C = \frac{12}{s} \frac{\frac{1}{s}}{20 + \frac{1}{s}} = \frac{12}{s(1 + 20s)} = \frac{12}{20} \frac{1}{s\left(s + \frac{1}{20}\right)} = \frac{3}{5} \frac{1}{s\left(s + \frac{1}{20}\right)},$$

$$= \frac{3}{5}\left(\frac{20}{s} + \frac{-20}{s + \frac{1}{20}}\right) = 12\left(\frac{1}{s} - \frac{1}{s + \frac{1}{20}}\right).$$

Looking up Table 23.2 to find the inverse Laplace transform, we have

$$v_C(t) = 12\left(1 - e^{-\frac{t}{20}}\right)u(t).$$

Here using $u(t)$ is just another way to say $t \geq 0$.

From $\frac{1}{s\left(s + \frac{1}{20}\right)}$ to get $\frac{20}{s} + \frac{-20}{s + \frac{1}{20}}$ is called **partial fraction expansion**, or partial
fraction decomposition. There is a short cut to perform it. The reason for doing
partial fraction expansion is that we cannot find $\frac{1}{s\left(s + \frac{1}{20}\right)}$ in Table 23.2; however, both $\frac{1}{s}$
and $\frac{1}{s + \frac{1}{20}}$ are in Table 23.2.

To do partial fraction expansion is to find k_1 and k_2 in

$$\frac{1}{s\left(s + \frac{1}{20}\right)} = \frac{k_1}{s} + \frac{k_2}{s + \frac{1}{20}}.$$

To find k_1 we multiply s on both sides of the above equation, obtaining

$$\frac{1}{\left(s + \frac{1}{20}\right)} = k_1 + \frac{sk_2}{s + \frac{1}{20}}.$$

Let $s \to 0$, we have $\frac{1}{\left(0 + \frac{1}{20}\right)} = k_1 + 0$, that is, $k_1 = 20$.

To find k_2 we multiply $\left(s + \frac{1}{20}\right)$ on both sides of the above equation, obtaining

$$\frac{1}{s} = \frac{\left(s + \frac{1}{20}\right)k_1}{s} + k_2.$$

Let $s \to -\frac{1}{20}$, we have $\frac{1}{\left(-\frac{1}{20}\right)} = 0 + k_2$, that is, $k_2 = -20$.

Once you are good at it, you can use a **cover-up method** as explained below. You start with writing down the equation:

$$\frac{1}{s\left(s + \frac{1}{20}\right)} = \frac{k_1}{s} + \frac{k_2}{s + \frac{1}{20}}.$$

and you want to find k_1. On the left-hand side of the equation, you cover up the denominator, s, of the k_1 term as

$$\frac{1}{s\left(s + \frac{1}{20}\right)} = \frac{k_1}{s} + \frac{k_2}{s + \frac{1}{20}}.$$

You realize that as $s \to 0$ the covered factor will be 0. Pay attention to the left-hand side, as if the covered part is not there anymore. Let $s \to 0$ for the rest of the left-hand side (uncovered part), and the limit is your k_1 value 20.

Next, you want to find k_2. You cover up the denominator of the k_2 term $\left(s + \frac{1}{20}\right)$ as

$$\frac{1}{s\left(s + \frac{1}{20}\right)} = \frac{k_1}{s} + \frac{k_2}{s + \frac{1}{20}}.$$

You realize that as $s \to -\frac{1}{20}$ the covered factor will be 0. Pay attention to the left-hand side, as if the covered part is not there anymore. Let $s \to -\frac{1}{20}$ for the rest of the left-hand side (uncovered part), and the limit is your k_1 value -20.

There are many ways to find the partial fraction expansion (or partial fraction decomposition). Another way is to sum up the expansion and compare the coefficients in the numerators on both sides, as shown below.

$$\frac{1}{s\left(s + \frac{1}{20}\right)} = \frac{k_1}{s} + \frac{k_2}{s + \frac{1}{20}} = \frac{k_1\left(s + \frac{1}{20}\right) + k_2 s}{s\left(s + \frac{1}{20}\right)} = \frac{(k_1 + k_2)s + \frac{1}{20}k_1}{s\left(s + \frac{1}{20}\right)}.$$

By comparing the coefficients on both sides of the equations in the numerators, we have

$$k_1 + k_2 = 0 \text{ and } \frac{1}{20}k_1 = 1.$$

We get the same answer as before: $k_1 = 20$ and $k_2 = -20$.

In fact, the method of comparing coefficients is more useful if the characteristic polynomial contains complex roots. The characteristic polynomial is simply the nominator polynomial with respect to s. Let us use one example to see it is done.

Example

Find the inverse Laplace transform of

$$Y(s) = \frac{1}{s(s^2 + 6s + 25)}.$$

Solution

The characteristic polynomial has one real root of $s = 0$ and two complex conjugate roots, because

$$s(s^2 + 6000s + 25 \times 10^6) = s[(s+3)^2 + 4^2].$$

In Table 23.2, we have two entries containing the form of $(s + a)^2 + \omega^2$ in the denominator. We must use both entries in the expansion.

Thus,

$$Y(s) = \frac{1}{s(s^2 + 6s + 25)} = \frac{k_1}{s} + k_2 \frac{\omega}{(s+a)^2 + \omega^2} + k_3 \frac{(s+a)}{(s+a)^2 + \omega^2},$$

$$= \frac{k_1}{s} + k_2 \frac{4}{(s+3)^2 + 4^2} + k_3 \frac{(s+3)}{(s+3)^2 + 4^2}.$$

The coefficient k_1 can be obtained by the cover-up method, and $k_1 = 1/25$. We have

$$Y(s) = \frac{1}{s(s^2 + 6s + 25)} = \frac{\frac{1}{25}}{s} + k_2 \frac{4}{(s+3)^2 + 4^2} + k_3 \frac{(s+3)}{(s+3)^2 + 4^2}.$$

Our next step is to sum up the right-hand side of the above equation, and then compare the coefficients.

$$\frac{1}{s(s^2 + 6s + 25)} = \frac{\frac{1}{25}}{s} + k_2 \frac{4}{(s+3)^2 + 4^2} + k_3 \frac{(s+3)}{(s+3)^2 + 4^2},$$

$$= \frac{\left(\frac{1}{25}\right)(s^2 + 6s + 25) + k_2 4s + k_3(s+3)s}{s(s^2 + 6s + 25)}.$$

Numerator coefficients of the s^2 term on both sides:

$$0 = \left(\frac{1}{25}\right) + k_3.$$

Numerator coefficients of the s^1 term on both sides:

$$0 = \left(\frac{6}{25}\right) + k_2 4 + k_3 3.$$

Numerator coefficients of the s^0 term on both sides:

$$1 = 1.$$

We obtain

$$k_3 = -\frac{1}{25},$$

and

$$k_2 = \frac{\frac{6}{25} - \frac{3}{25}}{-4} = -\frac{3}{100}.$$

According to Table 23.2, the inverse Laplace transform is obtained as

$$y(t) = \frac{1}{25} - \frac{3}{100} e^{-3t} \sin{(4t)} - \frac{1}{25} e^{-3t} \cos{(4t)}, \text{ for } t \geq 0.$$

As another example, if we are asked to find the inverse Laplace transform of

$$Y(s) = \frac{1}{s^2(s^2 + 6s + 25)},$$

its partial fraction exposition is in the form of

$$Y(s) = \frac{1}{s^2(s^2 + 6s + 25)} = \frac{k_1}{s} + \frac{k_2}{s^2} + k_3 \frac{4}{(s+3)^2 + 4^2} + k_4 \frac{(s+3)}{(s+3)^2 + 4^2}.$$

Do not forget the $\frac{k_1}{s}$ term because the lower-order terms are required during expansion.

The inverse Laplace transform of the newer $Y(s)$ is in the form of

$$y(t) = k_1 + k_2 t + k_3 e^{-3t} \sin{(4t)} + k_4 e^{-3t} \cos{(4t)}, \text{ for } t \geq 0.$$

Oh, wait a minute, we have not told you how the Laplace transform is defined and how Tables 23.1 and 23.2 are made yet. The Laplace transform of a function $f(t)$ is defined by an integral as shown in the expression below with a complex variable s.

$$F(s) = \int_0^{\infty} f(t)e^{-st}dt.$$

Everything in Tables 23.1 and 23.2 can be verified with this definition.

Notes

Using the Laplace transform, we are able to convert differential equations to algebraic equations. In the Laplace domain, Ohm's law works for inductors and capacitors again. KVL and KCL also work. One can setup the equations directly in the Laplace domain.

After solving the Laplace-domain algebraic equations, the solution is a function in the Laplace domain. In order to obtain a time-domain solution (which a function of time), we need to perform the inverse Laplace transform on the Laplace-domain solution.

To perform the inverse Laplace transform, the important step is to express the Laplace-domain function as a summation of the functions in the Laplace transform pair table (i.e., the right column of Table 23.2). This procedure is called the partial fraction decomposition, which can be achieved by the cover-up method or the coefficient comparison method.

Each term in the partial fraction expansion appears in the right column of Table 23.2. A time-domain expression is then readily obtained by using the corresponding terms in the left column.

Exercise Problems

Problem 23.1 Solve the following differential equation using the Laplace transform method.

$$x''(t) + 4x'(t) + 3x(t) = 5$$

with initial conditions $x'(0) = 1$ and $x(0) = 2$.

Problem 23.2 Use the Laplace transform method to solve for the circuit in Fig. P23.1. The initial voltage in the capacitor is 3 V.

Fig. P23.1

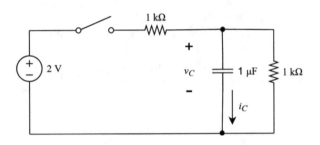

Problem 23.3 Use the Laplace transform method to solve for the circuit in Fig. P23.2. The initial current in the inductor is 3 A.

Fig. P23.2

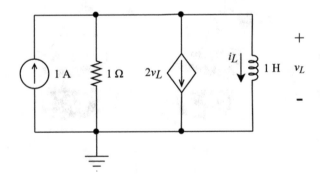

Solutions to Exercise problems are given in Book Appendix.

The Laplace transform is a powerful tool and you may have already seen it elsewhere, for example, in your Differential Equations class. By using the Laplace transform, we can do away with the differential equations. However, the Laplace transform has its drawbacks; it cannot handle the sources that are on the entire time as in steady-state analysis. By modifying the Laplace transform a little, more general sources can be accommodated.

Here comes the **Fourier transform** defined below for function $f(t)$, and we do NOT have the restriction of $t \geq 0$.

$$F(\omega) = \int_{-\infty}^{\infty} f(t)e^{-j\omega t}\,dt.$$

The biggest difference between the Fourier transform and the Laplace transform is the lower limit of the integral in the definition. If the function $f(t)$ is 0 for $t < 0$, then its Fourier transform and its Laplace transform are essentially the same, just replacing $j\omega$ by s.

For the Fourier transform, the counterparts of Tables 23.1 and 23.2 are Tables 24.1 and 24.2, respectively. Notice that the initial conditions no longer appear in the new tables. It is straightforward to verify that if the function $f(t)$ is an even function, then its Fourier transform is real and even. If the function $f(t)$ is an odd function, then its Fourier transform $F(\omega)$ is imaginary and odd.

In Table 24.2, $\text{sgn}(t)$ is the **signum function**, which is defined as

$$\text{sgn}(t) = \begin{cases} -1 & \text{for} \quad t < 0, \\ 0 & \text{for} \quad t = 0, \\ 1 & \text{for} \quad t > 0. \end{cases}$$

Table 24.1 Some properties of the Fourier transform

$f(t)$	$F(\omega) =$ Fourier transform of $f(t)$
$kf(t)$	$kF(\omega)$
$f_1(t) + f_2(t)$	$F_1(\omega) + F_2(\omega)$
$f'(t)$	$(j\omega)F(\omega)$
$f''(t)$	$(j\omega)^2 F(\omega)$
$f'''(t)$	$(j\omega)^3 F(\omega)$
$\int\limits_{-\infty}^{t} f(x)dx$	$\frac{1}{j\omega}F(\omega)$

Table 24.2 Fourier transform paris

$f(t)$	$F(\omega) =$ Fourier transform of $f(t)$
$\delta(t)$	1
1	$2\pi\delta(\omega)$
$u(t)$	$\pi\delta(\omega) + \frac{1}{j\omega}$
$\text{sgn}(t)$	$\frac{2}{j\omega}$
$e^{-at}u(t)$	$\frac{1}{s+j\omega}$
$e^{at}u(-t)$	$\frac{1}{s-j\omega}$
$e^{-a\mid t\mid}$	$\frac{2a}{a^2+\omega^2}$
$\sin(\omega_0 t)$	$j\pi[\delta(\omega + \omega_0) - \delta(\omega - \omega_0)]$
$\cos(\omega_0 t)$	$j\pi[\delta(\omega + \omega_0) + \delta(\omega - \omega_0)]$
$e^{j\omega_0 t}$	$2\pi\delta(\omega - \omega_0)$

Example

Use the Fourier transform to find $i_L(t)$ in the circuit shown in Fig. 24.1. The current source is sinusoidal $i_g(t) = 50\ cos\ (3t)$ A.

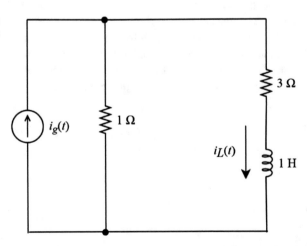

Fig. 24.1 An RL circuit with a sinusoidal current source for steady-state analysis

Solution

The Fourier domain equivalent of the circuit in Fig. 24.1 is shown in Fig. 24.2. According to Table 24.2,

$$I_g(\omega) = 50\pi[\delta(\omega + 3) + \delta(\omega - 3)].$$

Realizing that the circuit is a current divider, we have

$$I_L(\omega) = I_g(\omega)\frac{1}{1 + (3 + j\omega)},$$

$$= 50\pi[\delta(\omega + 3) + \delta(\omega - 3)]\frac{1}{4 + j\omega},$$

$$= 50\pi[\delta(\omega + 3) + \delta(\omega - 3)]\frac{(4 - j\omega)}{(4 + j\omega)(4 - j\omega)},$$

$$= 50\pi\frac{4 - j\omega}{16 + \omega^2}[\delta(\omega + 3) + \delta(\omega - 3)],$$

$$= 50\pi\frac{4 - j\omega}{16 + \omega^2}\delta(\omega + 3) + 50\pi\frac{4 - j\omega}{16 + \omega^2}\delta(\omega - 3),$$

$$= 50\pi\frac{4 - j(-3)}{16 + (-3)^2}\delta(\omega + 3) + 50\pi\frac{4 - j(3)}{16 + (3)^2}\delta(\omega - 3),$$

$$= 2\pi(4 + j3)\delta(\omega + 3) + 2\pi(4 - j3)\delta(\omega - 3),$$

$$= 8\pi[\delta(\omega + 3) + \delta(\omega - 3)] + 6j\pi[\delta(\omega + 3) - \delta(\omega - 3)].$$

Fig. 24.2 An RL circuit with a sinusoidal current source in Fourier domain representation

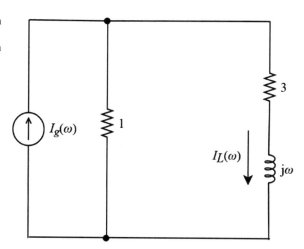

In the above calculation, we employed one unique property that is only valid for the delta function:

$$g(x)\delta(x - a) = g(a)\delta(x - a),\text{ for any continuous function } g(x).$$

Using Table 24.2 to find the inverse Fourier transform, we have the time domain expression:

$$i_L(t) = 8\cos(3t) + 6\sin(3t)\text{ A}.$$

In fact, the Fourier transform method is more powerful than the phasor method. The phasor method can only handle the steady-state sinusoid sources, while the Fourier transform method can handle any source input that is defined on $(-\infty, \infty)$.

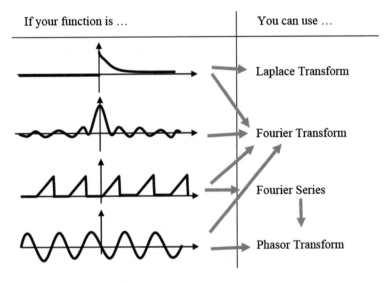

Notes

The Fourier transform is almost the same as the Laplace transform. The main difference is that the Laplace transform requires the time-domain functions defined in $t \geq 0$, while the Fourier transform does not have this restriction for the time-domain functions.

Ohm's law works for inductors and capacitors by using impedance. In many cases, we can use the relationship $s = j\omega$ to change the Laplace-domain equations to the Fourier-domain equations, and vice versa.

The Laplace transform can easily handle the initial conditions, while the Fourier transform can analyze steady-state functions.

Exercise Problems

Problem 24.1 Use the Fourier transform method to solve for the circuit in Fig. P24.1. The initial voltage in the capacitor is 3 V.

Fig. P24.1

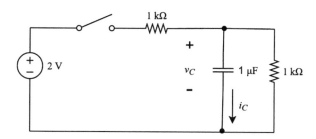

Problem 24.2 The input signal is a signum function. Find the inductor current i_L.

Problem 24.3 Repeat Problem 24.2 with $10 \cos (4t)$ being the input signal.

Fig. P24.2

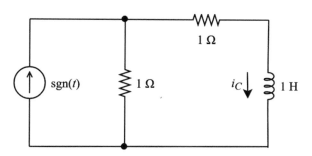

Solutions to Exercise problems are given in Book Appendix.

Second-Order Circuits

25

The voltage and current relationship in a **second-order circuit** is characterized by a second-order differential equation. A circuit consisting of a resistor (R), an inductor (L), and a capacitor (C) is most likely a second order circuit. We have seen an RLC circuit in action when we discussed steady-state analysis, in which the phasor notation was used, and the circuit order did not matter.

In this chapter, consider parallel and series RLC circuits with DC sources and switch actions. In other words, we are interested in the **step response** of the RLC circuits. This is not steady-state analysis, and differential equations may need to get involved. If we try to avoid the differential equations, we can use the Laplace transform or the Fourier transform.

The step response for a first-order system is fairly easy. It is an exponential transition from the initial value to the final value, with a time constant to determine the transition rate. For a second-order system, the transition is more complicated than an exponential function. The responses are categorized into three types (see illustrations in Fig. 25.1): the underdamped, critically damped, and **overdamped**.

© The Author(s), under exclusive license to Springer Nature Switzerland AG 2021
G. L. Zeng, M. Zeng, *Electric Circuits*,
https://doi.org/10.1007/978-3-030-60515-5_25

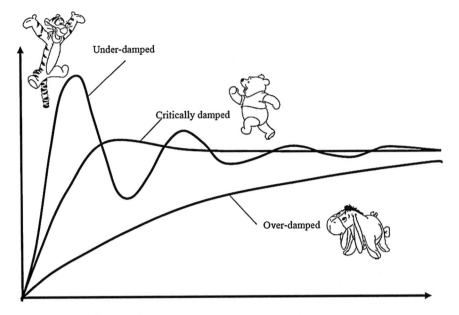

Fig. 25.1 The step response of a second-order system can have different shapes

Example
Find the step response $i(t)$ of the RLC circuit shown in Fig. 25.2. The initial conditions are zero. At $t = 0$, the voltage across the capacitor is zero, and the current in the inductor is zero.

Fig. 25.2 An RLC circuit

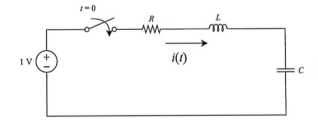

Solution
The initial conditions are zero. In other words, the system is initially relaxed. The voltage source together with the switch action can be treated as a unit step function. The Laplace transform of the source is $1/s$.

We will use the Laplace transform method to solve this problem. We first change the notation in Fig. 25.2 to Laplace domain notation as shown in Fig. 25.3. Using Ohm's law, we have

$$I(s) = \frac{\frac{1}{s}}{R + \frac{1}{sC} + sL} = \frac{1}{s} \frac{1}{R + \frac{1}{sC} + sL} = \frac{\frac{1}{L}}{s^2 + \frac{R}{L}s + \frac{1}{LC}}.$$

The denominator of $I(s)$ is a second-order polynomial; thus, this is a second-order circuit. There are three possible outcomes of $I(s)$:

1. [An underdamped solution]

$$I(s) = k_1 \frac{\omega}{(s+a)^2 + \omega^2} + k_2 \frac{s+a}{(s+a)^2 + \omega^2},$$

which corresponds to

$$i(t) = k_1 e^{-at} \sin(\omega t)u(t) + k_2 e^{-at} \cos(\omega t)u(t).$$

Since

$$I(s) = \frac{\frac{1}{L}}{s^2 + \frac{R}{L}s + \frac{1}{LC}}$$

and

$$I(s) = k_1 \frac{\omega}{(s+a)^2 + \omega^2} + k_2 \frac{s+a}{(s+a)^2 + \omega^2},$$

their nominators must be identical, that is,

$$\frac{1}{L} = k_1 \omega + k_2(s+a).$$

By comparing the coefficients, we must have

$$k_2 = 0 \quad \text{and} \quad k_1 = \frac{1}{L\omega}.$$

Thus,

$$i(t) = k_1 e^{-at} \sin(\omega t)u(t)$$

as shown in Fig. 25.4. In order to have an underdamped solution, the **characteristic equation** (setting the denominator to zero) $s^2 + \frac{R}{L}s + \frac{1}{LC} = 0$ must have two complex conjugate solutions: $s_{1,2} = -a \pm j\omega$. That is, we must have

$$\left(\frac{R}{L}\right)^2 - \frac{4}{LC} < 0$$

with $a = \frac{R}{2L}$ and $\omega = \sqrt{\frac{1}{LC} - \left(\frac{R}{2L}\right)^2}$.

2. [A critically damped solution]

$$I(s) = k_3 \frac{1}{s+a} + k_4 \frac{1}{(s+a)^2} = \frac{k_3(s+a) + k_4}{(s+a)^2},$$

which corresponds to

$$i(t) = k_3 e^{-at} u(t) + k_4 t e^{-at} u(t).$$

Since

$$I(s) = \frac{\frac{1}{L}}{s^2 + \frac{R}{L}s + \frac{1}{LC}}$$

and

$$I(s) = \frac{k_3(s+a) + k_4}{(s+a)^2},$$

their nominators must be identical, that is,

$$\frac{1}{L} = k_3(s+a) + k_4.$$

By comparing the coefficients, we must have

$$k_3 = 0 \quad \text{and} \quad k_4 = \frac{1}{L}.$$

Thus,

$$i(t) = \frac{1}{L} t e^{-at} u(t)$$

as shown in Fig. 25.4. In order to have a critically damped solution, the characteristic equation $s^2 + \frac{R}{L}s + \frac{1}{LC} = 0$ must have two identical real solutions: $s_{1,2} = -a$. That is, we must have

$$\left(\frac{R}{L}\right)^2 - \frac{4}{LC} = 0$$

with

$$a = \frac{R}{2L} = \frac{1}{\sqrt{LC}}.$$

3. [An over-damped solution]

$$I(s) = k_5 \frac{1}{s+a} + k_6 \frac{1}{s+b} = \frac{k_5(s+b) + k_6(s+a)}{(s+a)(s+b)},$$

which corresponds to

$$i(t) = k_5 e^{-at} u(t) + k_6 e^{-bt} u(t).$$

Since

$$I(s) = \frac{\frac{1}{L}}{s^2 + \frac{R}{L}s + \frac{1}{LC}}$$

and

$$I(s) = \frac{k_5(s+b) + k_6(s+a)}{(s+a)(s+b)},$$

their nominators must be identical, that is,

$$\frac{1}{L} = k_5(s+b) + k_6(s+a).$$

By comparing the coefficients, we must have

$$k_5 = -k_6 \text{ and } k_5 b + k_6 a = \frac{1}{L}.$$

They lead to

$$k_5(b-a) = \frac{1}{L},$$

$$k_5 = \frac{1}{L(b-a)},$$

$$k_6 = \frac{1}{L(a-b)}.$$

Thus,

$$i(t) = \frac{1}{L(b-a)} \left[e^{-at} - e^{-bt} \right] u(t)$$

as shown in Fig. 25.4. In order to have an over-damped solution, the characteristic equation $s^2 + \frac{R}{L}s + \frac{1}{LC} = 0$ must have two different real solutions: $s_1 = -a$ and $s_2 = -b$. That is, we must have

$$\left(\frac{R}{L}\right)^2 - \frac{4}{LC} > 0,$$

$$\text{with } a + b = \frac{R}{L} \quad \text{and} \quad ab = \frac{1}{LC}$$

or

$$a = \frac{\frac{R}{L} + \sqrt{\left(\frac{R}{L}\right)^2 - \frac{4}{LC}}}{2} \quad \text{and} \quad b = \frac{\frac{R}{L} - \sqrt{\left(\frac{R}{L}\right)^2 - \frac{4}{LC}}}{2}.$$

Fig. 25.3 An RLC circuit using Laplace transform notation

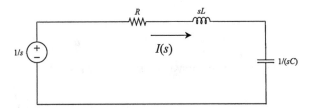

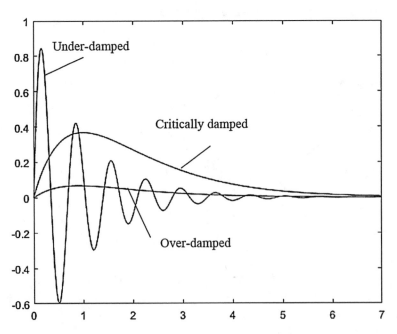

Fig. 25.4 Three potential outcomes of the current in the circuit of Fig. 25.2 or Fig. 25.5

Finally, we consider a natural response of a circuit. A natural response is the response due to the initial conditions. There are no external sources.

Example

Consider the RLC circuit shown in Fig. 25.5. The initial conditions are $i(0) = 0.5$ A and $v(0) = 0$ V. Find the natural response $v(t)$ for $t \geq 0$.

Fig. 25.5 An RLC circuit with initial current $i(0) = 0.5$ A and initial voltage $v(0) = 0$ V

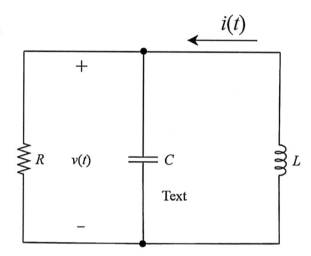

Solution

In Table 1 we learned that the Laplace transform of $\frac{di(t)}{dt}$ is $sI(s) - i(0)$. For the inductor, we have $v(t) = L\frac{di(t)}{dt}$. Thus, in the Laplace domain, this relationship is $V(s) = LsI(s) - Li(0)$. This implies that if there is a non-zero initial condition, we must add a source $-Li(0)$ to the circuit. The relationship $V(s) = LsI(s) - Li(0)$ also implies that $-Li(0)$ is an independent voltage source. Therefore, we can obtain a Laplace domain equivalent circuit as in Fig. 25.6.

If we compare the voltage sources in Figs. 25.3 and 25.6, they are different. The source in Fig. 25.3 is $1/s$ and in the time domain it is a step function according to Table 2 in Chap. 23. A step function power source keeps providing energy to the circuit once it is on. On the other hand, the source in Fig. 25.6 is $Li(0)$ (without the $1/s$ part) and in the time domain it is a delta function according to Table 2 in Chap. 23. A delta function power source only provides energy to the circuit at one moment and provides no energy to the circuit after that.

When using the Laplace transform approach, the initial conditions are converted into dependent sources, and then the circuits are assumed to have zero initial conditions.

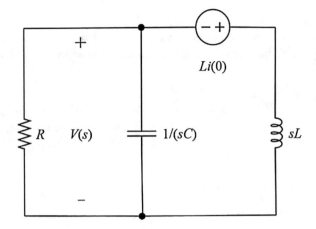

We can now set up a node-voltage equation as

$$\frac{V(s)}{R} + sCV(s) + \frac{V(s) + Li(0)}{sL} = 0,$$

which leads to

$$V(s) = \frac{\frac{i(0)}{s}}{\frac{1}{R} + sC + \frac{1}{sL}} = \frac{i(0)}{\frac{1}{R}s + Cs^2 + \frac{1}{L}} = \frac{i(0)}{C}\frac{1}{s^2 + \frac{1}{RC}s + \frac{1}{LC}}.$$

The characteristic equation is

$$s^2 + \frac{1}{RC}s + \frac{1}{LC} = 0,$$

which can have three different types of solutions depending on the values of the R, L, and C.

1. Two complex conjugate solutions, when $\frac{1}{(RC)^2} < \frac{4}{LC}$. [Underdamped]

   The solution is in the form of

   $$v(t) = k_1 e^{-at} \sin(\omega t)u(t) + k_2 e^{-at} \cos(\omega t)u(t).$$

   Since

   $$V(s) = \frac{i(0)}{C}\frac{1}{s^2 + \frac{1}{RC}s + \frac{1}{LC}}$$

   and

$$V(s) = k_1 \frac{\omega}{(s+a)^2 + \omega^2} + k_2 \frac{s+a}{(s+a)^2 + \omega^2},$$

their nominators must be identical, that is,

$$\frac{i(0)}{C} = k_1\omega + k_2(s+a).$$

By comparing the coefficients, we must have

$$k_2 = 0 \text{ and } k_1 = \frac{i(0)}{C\omega}.$$

Thus,

$$v(t) = \frac{i(0)}{C\omega} e^{-at} \sin(\omega t) u(t)$$

as shown in Figure 25.4.
2. Two identical real solutions, when $\frac{1}{(RC)^2} = \frac{4}{LC}$. [Critically damped]

The solution is in the form of

$$v(t) = k_3 e^{-at} u(t) + k_4 t e^{-at} u(t).$$

Since

$$V(s) = \frac{i(0)}{C} \frac{1}{s^2 + \frac{1}{RC}s + \frac{1}{LC}}$$

and

$$V(s) = k_3 \frac{1}{s+a} + k_4 \frac{1}{(s+a)^2} = \frac{k_3(s+a) + k_4}{(s+a)^2},$$

their nominators must be identical, that is,

$$\frac{i(0)}{C} = k_3(s+a) + k_4.$$

By comparing the coefficients, we must have

$$k_3 = 0 \text{ and } k_4 = \frac{i(0)}{C}.$$

Thus,

$$v(t) = \frac{i(0)}{C} t e^{-at} u(t)$$

as shown in Fig. 25.4.

3. Two different real solutions, when $\frac{1}{(RC)^2} > \frac{4}{LC}$. [Over-damped]

The solution is in the form of

$$v(t) = k_5 e^{-at} u(t) + k_6 e^{-bt} u(t).$$

Since

$$V(s) = \frac{i(0)}{C} \frac{1}{s^2 + \frac{1}{RC} s + \frac{1}{LC}}$$

and

$$V(s) = \frac{k_5(s+b) + k_6(s+a)}{(s+a)(s+b)},$$

their nominators must be identical, that is,

$$\frac{i(0)}{C} = k_5(s+b) + k_6(s+a).$$

By comparing the coefficients, we must have

$$k_5 = -k_6 \text{ and } k_5 b + k_6 a = \frac{i(0)}{C}.$$

They lead to

$$k_5(b - a) = \frac{i(0)}{C},$$

$$k_5 = \frac{i(0)}{C(b - a)},$$

$$k_6 = \frac{i(0)}{C(a - b)}.$$

Thus,

$$v(t) = \frac{i(0)}{C(b - a)} \left[e^{-at} - e^{-bt} \right] u(t)$$

as shown in Fig. 25.4.

You may have noticed that sometimes we use the notation $f(0)$ and other times we use the notation $f(0^-)$, for example, in Table 1 of Chap. 23. Are they the same? If the function $f(t)$ is not allowed to have a sudden change of values such as the inductor current or capacitor voltage, we have $f(0^-) = f(0) = f(0^+)$, and we can use either $f(0^-)$ or $f(0)$ to indicate the initial condition. In general, use $f(0^-)$ for the initial condition if you are not sure whether $f(0^-)$ and $f(0^+)$ are the same.

Notes
A typical second-order circuit consists of RLC and is described by a second-order differential equation. The response to a simple switch action can be underdamped, critically damped, or over-damped, depending on the values of R, L, and C.

The superposition principle allows us to study the natural response and step response. For the natural response, the initial values are treated as delta function power sources. In the step response studies, the initial conditions are assumed to be zero.

The Laplace transform is effective in these transient response studies.

Exercise Problems

Problem 25.1 In a series RLC circuit, what does the step response look like when $R = 0$? The input is step voltage source, and the output is the voltage across the capacitor.

Fig. P25.1

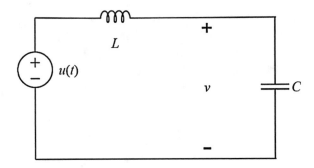

Problem 25.2 In a parallel RLC circuit, what does the step response look like when $R = \infty$? The input is step current source, and the output is the current through the inductor.

Fig. P25.2

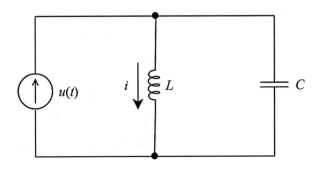

Problem 25.3 The circuit in Fig. P25.3 is neither a series nor a parallel circuit. It is still a second-order RLC circuit. Find the conditions for circuit to be underdamped, critically damped and over-damped.

Fig. P25.3

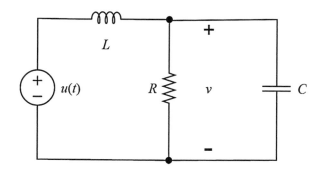

Solutions to Exercise problems are given in Book Appendix.

Filters

26

This chapter revisits sinusoidal steady-state analysis, emphasizing the frequency dependency. In Chap. 19, the emphasis was on a fixed given frequency; the frequency stays the same throughout the entire circuit.

This chapter assumes that the circuit has only one steady-state sinusoidal voltage source and its frequency can be anywhere from 0 to ∞. The output of the circuit is a voltage. We are interested in the ratio of the output voltage over the input voltage as a function of the frequency ω. We are more interested in the magnitude (i.e., the norm) of the ratio, which is referred to as the gain. The gain is the same, whether the circuit is represented in time domain, Fourier domain, or phasor notation.

We can either use the phasor method or the Fourier transform method to find the output voltage. Actually, these two approaches are the same for filter design and analysis. To get started, let us consider a simple RC circuit in Fig. 26.1. This is a voltage divider, and the gain is readily expressed as

$$\text{gain}(\omega) = \left| \frac{V_{out}(\omega)}{V_{in}(\omega)} \right| = \left| \frac{\frac{1}{j\omega C}}{R + \frac{1}{j\omega C}} \right| = \frac{1}{|j\omega RC + 1|} = \frac{1}{\sqrt{(\omega RC)^2 + 1}}.$$

Let us look at two extreme cases of $\omega = 0$ and $\omega \to \infty$, respectively. When $\omega = 0$, the gain reaches its maximum value of 1. When $\omega \to \infty$, the gain approaches its maximum value of 0. The circuit acts like a **lowpass filter**, having a higher gain at low frequencies and lower gain at high frequencies (see Fig. 26.2).

© The Author(s), under exclusive license to Springer Nature Switzerland AG 2021
G. L. Zeng, M. Zeng, *Electric Circuits*,
https://doi.org/10.1007/978-3-030-60515-5_26

Fig. 26.1 An RC circuit represented in the Fourier domain

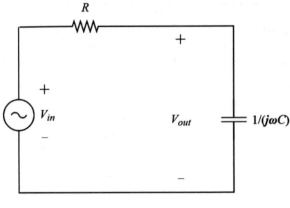

Fig. 26.2 Frequency response of a low-pass filter

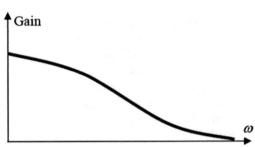

If we switch the positions of the capacitor and the resistor in Fig. 26.1, the new circuit (see Fig. 26.3) will behave like a **highpass filter** (see Fig. 26.4), in the sense that the gain is higher at high frequencies and the gain is lower at low frequencies. In fact, the circuit in Fig. 26.3 is voltage divider, and its gain is evaluated as

Fig. 26.3 An RC circuit represented in the Fourier domain

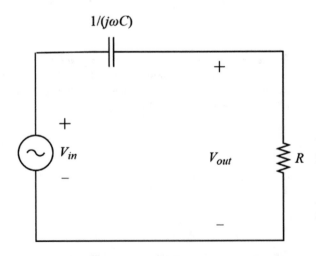

Fig. 26.4 Frequency
response of a high-pass filter

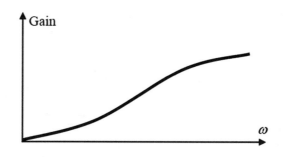

$$\text{gain}(\omega) = \left| \frac{V_{\text{out}}(\omega)}{V_{\text{in}}(\omega)} \right| = \left| \frac{R}{R + \frac{1}{j\omega C}} \right| = \frac{|j\omega RC|}{|j\omega RC + 1|} = \frac{\omega RC}{\sqrt{(\omega RC)^2 + 1}}.$$

When $\omega = 0$, the gain reaches its minimum value of 0. When $\omega \to \infty$, the gain approaches its maximum value of 1.

Circuits that specifically attenuate some frequencies and enhance other frequencies are called **filters**. Filters consisting of resistors, capacitors, and inductors are **passive filters** because they do not require an external power source (beyond the input signal). **Active filters**, on the other hand, require external power source to operate. Most active filters use op-amps. An inverting amplifier is shown in Fig. 26.5, and its gain is given as

$$\text{gain} = \left| \frac{v_{\text{out}}}{v_{\text{in}}} \right| = \left| -\frac{R_f}{R_i} \right| = \frac{R_f}{R_i}.$$

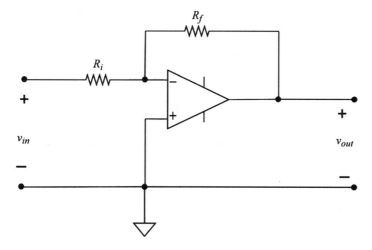

Fig. 26.5 An inverting amplifier using resistors

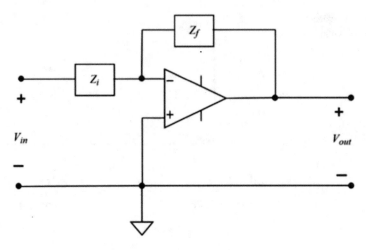

Fig. 26.6   A general **inverting amplifier**

If we replace the resistors in Fig. 26.5 by general impedances as shown in Fig. 26.6 with the gain

$$\text{gain} = \left| \frac{V_{\text{out}}}{V_{\text{in}}} \right| = \left| -\frac{Z_f}{Z_i} \right|,$$

we can obtain many useful active filters. One special case is shown in Fig. 26.7, and its gain is

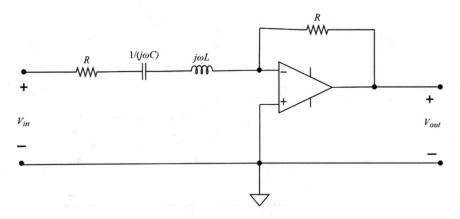

Fig. 26.7   A bandpass filter

Fig. 26.8 Frequency
response of a bandpass filter

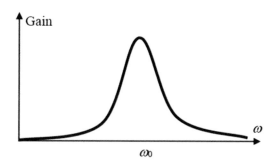

$$\text{gain} = \left|\frac{V_{\text{out}}}{V_{\text{in}}}\right| = \left|-\frac{R}{R + j\omega L + \frac{1}{j\omega C}}\right| = \left|-\frac{R}{R + j\left(\omega L - \frac{1}{\omega C}\right)}\right| = \frac{R}{\sqrt{R^2 + \left(\omega L - \frac{1}{\omega C}\right)^2}}.$$

When $\omega L = \frac{1}{\omega C}$ or $\omega = \frac{1}{\sqrt{LC}}$, the gain reaches to its maximum value of 1. When $\omega \to 0$ or $\omega \to \infty$, the gain tends to its minimum value of 0. Therefore, this is a **bandpass filter** (see Fig. 26.4). The frequency $\omega_0 = \frac{1}{\sqrt{LC}}$ is called the **resonance radian frequency**, at which the effects of a capacitor and an inductor cancel out (they can be replaced by a short circuit). At the resonance frequency, the circuits in Figs. 26.1 and 26.3 are the same with $R_f = R_i = R$ (Fig. 26.8).

Likewise, the circuit shown in Fig. 26.9 is another bandpass filter configuration with the gain calculated as

$$\text{gain} = \left|\frac{V_{\text{out}}}{V_{\text{in}}}\right| = \left|-\frac{\frac{1}{R} + \frac{1}{j\omega L} + j\omega C}{R}\right| = \left|-\frac{1}{1 + jR\left(\omega C - \frac{1}{\omega L}\right)}\right| = \frac{1}{\sqrt{1 + R^2\left(\omega C - \frac{1}{\omega L}\right)^2}}.$$

At the frequency $\omega_0 = \frac{1}{\sqrt{LC}}$, the effects of a capacitor and an inductor cancel out (they can be replaced by an open circuit).

Finally, we would like to mention that people usually express the filter gain in "**dB**" (**decibel**), which is defined by 20 times the common logarithm (that is, the logarithm to base 10) as

$$\text{Number of decibels} = 20 \log_{10} \left|\frac{v_{\text{out}}}{v_{\text{in}}}\right|.$$

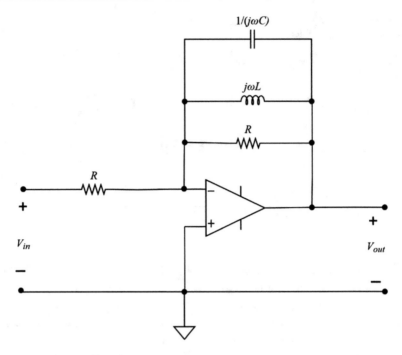

Fig. 26.9 A bandpass filter

For example, a voltage gain of 1 is 0 dB; a voltage gain of 10 is 20 dB; a voltage gain of 100 is 40 dB. If the voltage gain is less than 1, the dB value is negative. For example, the voltage gain of 0.1 corresponds to −20 dB.

Notes

In this chapter, the focus is the circuit gain as a function of the frequency. A circuit that intentionally enhances some frequencies and attenuates other frequencies is called a filter. To study the filter behavior is in the category of sinusoidal steady-state analysis. The Fourier transform (or equivalently, the phasor transform) is a natural fit for this type of studies.

It is interesting to notice that for a second-order RLC filter, there is a certain frequency, at which the effects of the capacitor and inductor cancel each other out. This frequency is called the resonant frequency of the filter.

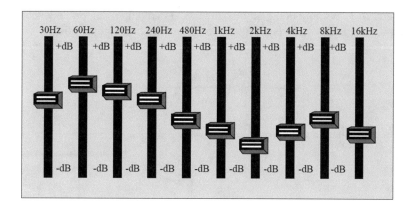

Exercise Problems

Problem 26.1 Without doing any mathematical derivation, determine whether the Sallen-Key filter shown in Fig. P26.1 a lowpass filter or a highpass filter.

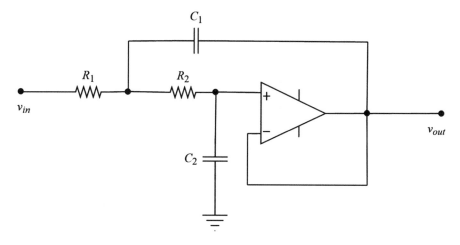

Fig. P26.1

Problem 26.2 Without doing any mathematical derivation, determine whether the Sallen-Key filter shown in Fig. P26.2 a lowpass filter or a highpass filter.

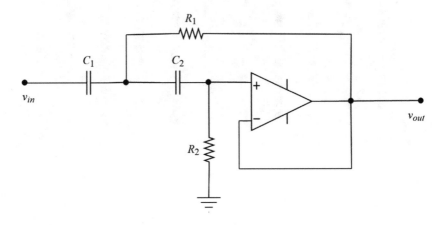

Fig. P26.2

Problem 26.3 Without doing any mathematical derivation, determine whether the Sallen-Key filter shown in Fig. P26.3 a lowpass filter or a highpass filter or none of them.

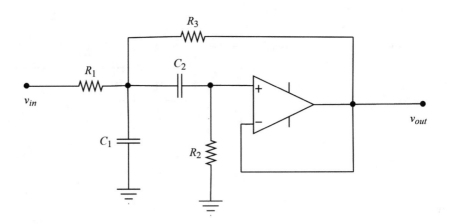

Fig. P26.3

Solutions to Exercise problems are given in Book Appendix.

Wrapping Up

27

Ohm's law is the foundation of electrical engineering. The resistors obey Ohm's law. We can use KCL and KVL to set up linear equations and solve for the unknowns. Circuits can be simplified by using Thévenin equivalent and Norton equivalent circuits. If there is a dependent source in the circuit, we always need an equation to describe that dependent source. An op-amp is a dependent source. An ideal op-amp is a good approximation of a practical op-amp. For a stable op amp circuit, the feedback path should go to the inverting input labeled by a "−" sign.

You do not have to solve a circuit problem by solving system of linear equations. Many times, a simple voltage divider or current divider method can give you the answer quickly.

Everything goes well until capacitors and inductors show up. The voltage and current for a capacitor or an inductor are related with a derivative or an integral, and Ohm's law is not obeyed. The KCL and KVL equations become differential equations. We admit that it is not easy to work with differential equations. This is where the various transform methods come in. The sole purpose of these methods is to make Ohm's law work again for capacitors and inductors so that the differential equations reduce to algebraic equations.

The Laplace transform can be used when we are considering initial conditions and switch actions. In the Laplace domain, capacitors follow Ohm's Law $V_C = \left(\frac{1}{sC}\right)I_C$ and inductors follow Ohm's Law $V_L = (sL)I_L$. The Fourier transform is useful for steady-state analysis and frequency response studies. If the source is pure sinusoidal, the phasor method is a method dedicated to evaluating the magnitude response and phase shift. In the Fourier domain and the phasor method, capacitors follow Ohm's Law $V_C = \left(\frac{1}{j\omega C}\right)I_C$ and inductors follow Ohm's Law $V_L = (j\omega L)I_L$. Basically, you either use "s" or use "$j\omega$" for the time-domain derivative. Filter analysis is easier to perform by using "$j\omega$."

Every periodic function can be expanded into a Fourier series. Each term in the expansion is sinusoidal, and thus the phasor method can be applied to each term in the Fourier series. The superposition principle can be used to obtain the result if the

© The Author(s), under exclusive license to Springer Nature Switzerland AG 2021
G. L. Zeng, M. Zeng, *Electric Circuits*,
https://doi.org/10.1007/978-3-030-60515-5_27

input is a general periodic function. In fact, the Fourier transform and the Fourier series are closely related. The Fourier transform is defined for non-periodic functions, and the Fourier series for periodic functions. The Fourier transform is the limiting case for the Fourier series by letting the period $T \to \infty$.

For a first-order circuit, you can just write down the time-domain step response of a capacitor voltage or an inductor current without setting up equations. The voltage across a capacitor and the current in an inductor cannot have a sudden jump. They can change slowly from the initial value to the final value along an exponential path, governing by a time constant.

The time-domain step response of a second-order circuit also takes time from the initial value to the final value but following one of the three types of paths. One type is called the "underdamped" in which the response curve oscillates up-and-down many times before settling down as combined $e^{-at} \cos(\omega t)$ and $e^{-at} \sin(\omega t)$ functions. Another type is called the "overdamped" in which the response curve has the least or no oscillations at all; it is a combination of two exponential functions e^{-at} and e^{-bt} functions. The third type is called the "critically damped," and the response is a combination of an e^{-at} curve and a te^{-at} curve.

Any circuit can be a filter. A way to describe a filter is by using a gain function. The gain is the magnitude of the transfer function. The **transfer function** is the ratio of the output over the input in the Fourier or Laplace domain.

> **Notes**
> Circuit analysis requires solving KVL and KCL equations. When the transform methods are used, Ohm's law works for inductors, capacitors, and resistors. The KVL and KCL equations are now algebraic equations. The transform-domain solution needs to be converted back to the time domain. This step is usually achieved by the partial fraction decomposition and by table lookup.

Exercise Problems

Problem 27.1 Are KVL equations and mesh equations the same? Are KCL equations and node equations the same.

Problem 27.2 Do you have a preference regarding to nodes equations or mesh equations?

Problem 27.3 The main purpose of the Laplace transform and the Fourier transform is to avoid solving differential equations in the time domain. Why do we need both the Laplace transform and the Fourier transform? Can we just learn one of them?

Problem 27.4 Which is more powerful, the Fourier transform method or the phasor method?

Problem 27.5 Name one most important concept in electric circuits.

Solutions to Exercise problems are given in Book Appendix.

Appendix. Solutions to Exercise Problems

Chapter 1. Voltage, Current, and Resistance

Problem 1.1 Either of the following two symbols represents a DC voltage source. Here "V" is an abbreviation of "Volts." "Volt" is a unit of voltage.

Fig. P1.1a

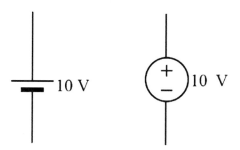

Determine whether the following configurations of voltage sources are valid or invalid. Why?

Solution
(a) Valid. Two (or more) DC voltage sources can be connected in parallel. There is no current flowing in the circuit.
(b) Invalid. These two voltage sources with different voltages cannot be connected in parallel.
(c) Valid. Voltage sources can be connected in series. The total voltage is the sum of each voltage source. V_{total} in this case is 4 V.
(d) Valid. Voltage sources can be connected in series, even though the source polarities are not consistent. The opposite polarity results in a negative voltage. The total voltage in this case is 2 V.

G. L. Zeng, M. Zeng, *Electric Circuits*, https://doi.org/10.1007/978-3-030-60515-5

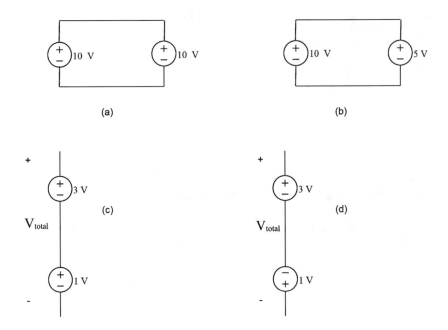

(a)   (b)

(c)   (d)

Fig. P1.1b

Problem 1.2 The purpose of a voltage source in a circuit is to cause the current to flow in a circuit. The flow of the electric current can be converted into something useful to us. For example, the electric current running through a heating wire can generate heat. The electric current running through a light bulb creates light. The electric current running through an electric motor causes motion. Please comment on the circuit shown whether this circuit is useful.

Fig. P1.2

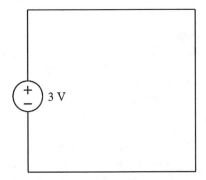

Solution

This circuit should never be allowed because voltage source is shorted. One should never ever short a voltage source!

This situation is similar to Case (b) in Problem 1.1, the above circuit is equivalent to two voltage sources connected in parallel, one with a voltage of 3 V and the other with a voltage of 0 V.

Fig. S1.2

The voltage sources in the textbooks are ideal, in the sense that their voltages keep constant regardless the rest of the circuit. In our everyday life, batteries and power supplies are close approximations of the ideal voltage sources. DO NOT short them! Otherwise, the fuse of the power supplies may be blown. The house circuit breaker may be trigged. Fire may be caused.

Problem 1.3 Even though we do not see them in everyday life, there are such things called "current sources." The ideal current source provides constant current, regardless the rest of the circuit. The symbol for a current source is shown below. Here "A" is an abbreviation of "Amperes." "Amperes" is a unit of current.

Fig. P1.3a

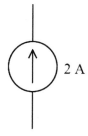

Determine whether the following configurations of current sources are valid or invalid. Why?

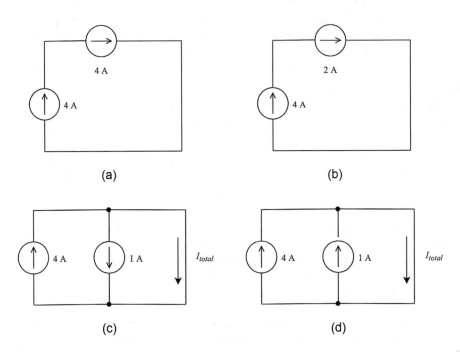

(a)   (b)

(c)   (d)

Fig. P1.3b

Solution

(a) Valid. Two or more current sources can be connected in series as long as their current values are the same.

(b) Invalid. If the current values are not the same, they cannot be connected in series. They provide conflicting current values.

(c) Valid. Current sources can be connected in parallel. Since these two current sources are connected with opposite polarity, the total current to the difference of the currents provided by the two current sources. This this case, the total current is 3 A.

(d) Valid. The total current is the sum of the two sources when their polarities are the same. In this case, the total current is 5 A.

Problem 1.4 Determine whether the following circuits are valid.

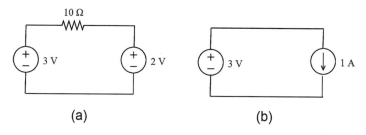

(a)   (b)

Fig. P1.4

Solution
(a) Valid. This circuit is equivalent to the circuit shown below. This circuit is different from that in Problem 1.1(b), in which there is no resistor in the circuit. The resistor here is labeled with "10 Ω." The unit of resistance is Ohm (Ω).

Fig. S1.4

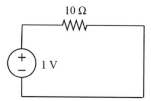

(b) Valid. The voltage source maintains its voltage regardless the current drawn from it and it allows be connected to a current source. Likewise, the current source maintains is current output regardless the voltage.

Problem 1.5 Draw a schematic for the flashlight circuit.

Solution
The schematic for the flashlight is shown below. The light bulb acts as a resistor.

Fig. S1.5

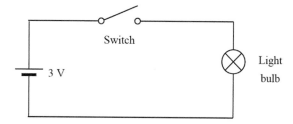

Chapter 2. DC Power Supply and Multimeters

Problem 2.1 You are given a power supply and a circuit schematic shown. Suggest three ways to connect the power supply to the 1 kΩ resistor.

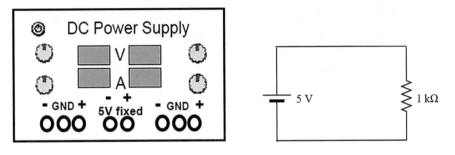

Fig. P2.1

Solution

The three ways are illustrated below:

(a)

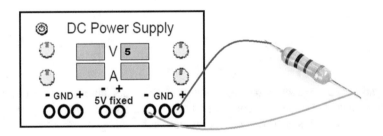

Fig. S2.1a

(b)

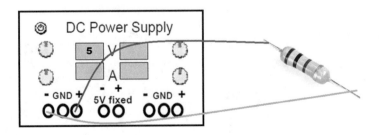

Fig. S2.1b

(c)

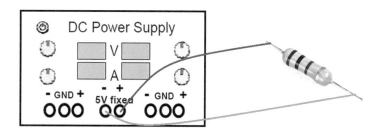

Fig. S2.1c

Problem 2.2 Identify the mistakes in using a multimeter.

(a) Trying to measure the voltage across the power supply.

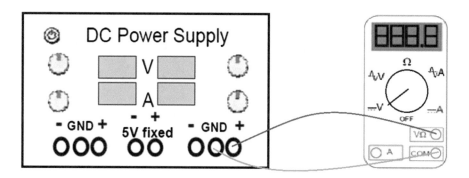

Fig. P2.2a

(b) Trying to measure the current through the resistor.

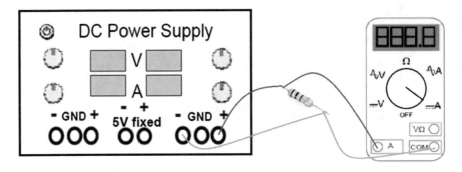

Fig. P2.2b

(c) Trying to measure the resistance of the resistor.

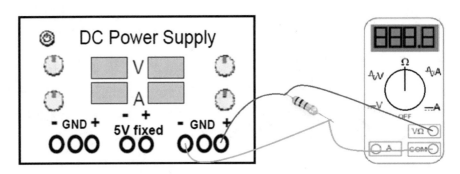

Fig. P2.2c

Solution

(a) In a power supply, the GND (ground) terminal is the chassis ground terminal. Depending on the application, this chassis ground may be connected to the negative (−) terminal, may be connected to the positive (+) terminal, or may not be connected to any terminals (floating). The DC power output is through the positive (+) and negative (−) terminals. A correct connection to measure the output voltage is shown in the following figure.

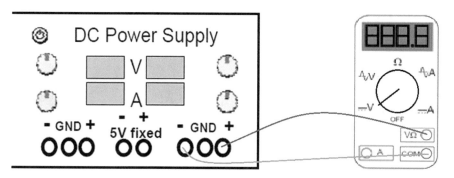

Fig. S2.2a

(b) Never ever do this when measure current. This will burn the multimeter or its fuse. The ammeter has very low resistance. When an ammeter is directly connected to the power supply output, a huge current will run through the meter. In order to measure the current through a resistor, the ammeter must be connected in series with the resistor as shown below.

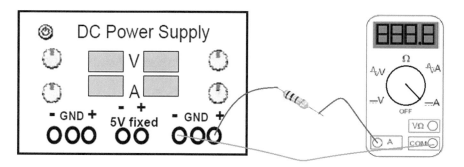

Fig. S2.2b

(c) When measuring the resistance of a resistor, we must remove the resistor from the circuit and measure the resistor by itself as shown below for two reasons. The first reason is that the resistor may be connected to other components in the circuit is a rather complicated way. For example, this resistor may be connected to a smaller resistor in parallel. The reading of the measurement will be affected by other components connected to your resistor of interest. The second reason is that in the resistance measuring mode, the ohm meter is actually measuring current while using meter's internal power to supply the voltage. If the circuit's power is on, the current running through the resistor will affect the resistance reading.

Fig. S2.2c

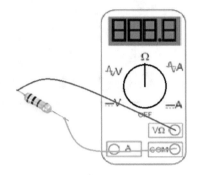

Chapter 3. Ohm's Law

Problem 3.1 Use Ohm's law to calculate the current in the circuit.

Solution

(a) The voltage source polarity is consistent with the current direction, and thus

$$I = \frac{V}{R} = \frac{10 \text{ V}}{5 \, \Omega} = 2 \text{ A}.$$

The actual current is running from left-to-right through the 5 Ω resistor.

(b) The voltage source polarity is consistent with the current direction, but the voltage source has a negative value. We thus have

$$I = \frac{V}{R} = \frac{-10 \text{ V}}{5 \, \Omega} = -2 \text{ A}.$$

Therefore, the actual current is running from right-to-left through the 5 Ω resistor. The negative sign in the answer "−2 A" implies that the current is running in the opposite direction as that indicated by the arrow in the figure.

(c) The voltage source polarity is inconsistent with the current direction, and thus we have

$$I = \frac{-V}{R} = \frac{-(10 \text{ V})}{5 \, \Omega} = -2 \text{ A}.$$

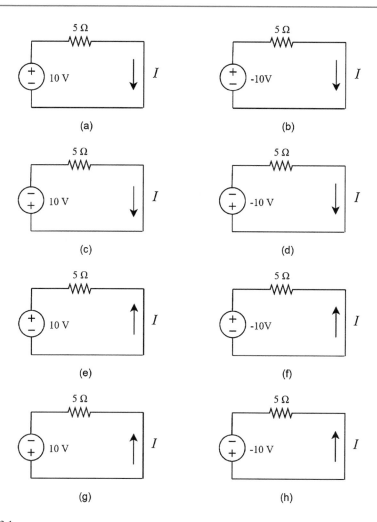

Fig. P3.1

The negative sign in the answer "−2 A" implies that the current is running in the opposite direction as that indicated by the arrow in the figure.

(d) The voltage source polarity is inconsistent with the current direction, and the voltage value of the voltage source is negative. We now have a case of "double negative." Thus, we have

$$I = \frac{-V}{R} = \frac{-(-10\ \text{V})}{5\ \Omega} = 2\ \text{A}.$$

The positive answer "2 A" implies that the current is running in the direction as that indicated by the arrow in the figure.

(e) The voltage source polarity is inconsistent with the current direction, and thus we have

$$I = \frac{-V}{R} = \frac{-(10 \text{ V})}{5 \text{ }\Omega} = -2 \text{ A}.$$

The negative sign in the answer "-2 A" implies that the current is running in the opposite direction as that indicated by the arrow in the figure. The actual current is running from left-to-right through the 5 Ω resistor.

(f) The voltage source polarity is inconsistent with the current direction, and the voltage value of the voltage source is negative. We now have another case of "double negative." Thus, we have

$$I = \frac{-V}{R} = \frac{-(-10 \text{ V})}{5 \text{ }\Omega} = 2 \text{ A}.$$

The positive answer "2 A" implies that the current is running in the direction as that indicated by the arrow in the figure. The current is running from right-to-left through the 5 Ω resistor.

(g) The voltage source polarity is consistent with the current direction, and thus

$$I = \frac{V}{R} = \frac{10 \text{ V}}{5 \text{ }\Omega} = 2 \text{ A}.$$

The actual current is running from right-to-left through the 5 Ω resistor.

(h) The voltage source polarity is consistent with the current direction, but the voltage source value is negative. Thus,

$$I = \frac{V}{R} = \frac{-10 \text{ V}}{5 \text{ }\Omega} = -2 \text{ A}.$$

The actual current is running from left-to-right through the 5 Ω resistor.

Problem 3.2 According to the partial circuit shown, use Ohm's law to calculate the voltage across the resistor. You must use the voltage polarity and current direction specified in the figure.

Solution
Ohm's law can only be applied to a resistor. The "voltage V" in Ohm's law is always the voltage across the resistor. The "current I" in Ohm's law is always the current through the resistor. The relationship $V = IR$ holds when the voltage polarity and the current direction are consistent: The current arrow points from "+" to "−" of the voltage labels on the two ends of the resistor.

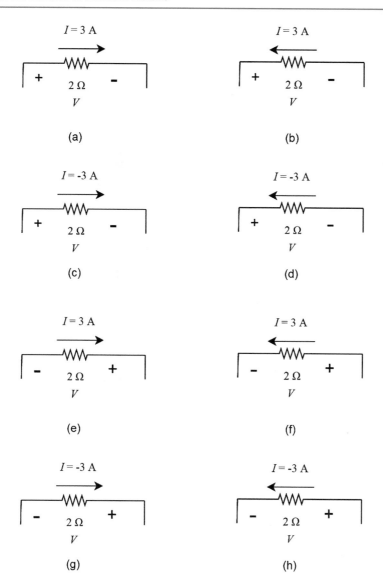

Fig. P3.2

(a) The notation is consistent, and the current is positive. Thus,

$$V = IR = (3 \text{ A})(2 \text{ } \Omega) = 6 \text{ V}.$$

(b) The notation is inconsistent, and a negative sign is required to compensate for this inconsistency. Thus,

$$V = -IR = -(3 \text{ A})(2 \, \Omega) = -6 \text{ V}.$$

The voltage is negative, which means that the right end of the resistor has a higher electric potential than the potential at the left end of the resistor.

(c) The notation is consistent, but the current is negative. We have

$$V = IR = (-3 \text{ A})(2 \, \Omega) = -6 \text{ V}.$$

The voltage is negative, which means that the right end of the resistor has a higher electric potential than the potential at the left end of the resistor. The current is actually running from right-to-left through the 2 Ω resistor.

(d) The notation is inconsistent, and the current is negative. We have a "double-negative" case,

$$V = -IR = -(-3 \text{ A})(2 \, \Omega) = 6 \text{ V}.$$

The voltage is positive, which means that the left end of the resistor has a higher electric potential than the potential at the right end of the resistor.

(e) The notation is inconsistent, and a negative sign is required for Ohm's law. We have

$$V = -IR = -(3 \text{ A})(2 \, \Omega) = -6 \text{ V}.$$

The voltage is higher on the left. The actual current always runs from high to low in a resistor.

(f) The notation is consistent. We have

$$V = IR = (3 \text{ A})(2\Omega) = 6 \text{ V}.$$

(g) The notation is inconsistent, and the current is negative. We have

$$V = -IR = (-3 \text{ A})(2\Omega) = 6 \text{ V}.$$

(h) The notation is consistent, but the current is negative. Thus,

$$V = IR = (-3 \text{ A})(2 \, \Omega) = -6 \text{ V}.$$

Problem 3.3 You are given an electrical element without any labels. You connect the element with a variable voltage source. You make some voltage/current measurements as shown in the table. What most likely is this element?

V (Volts)	I (Amperes)
10	5
0	0
−10	−5

Fig. P3.3

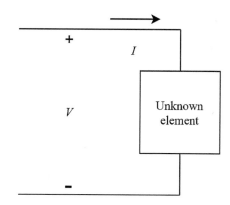

Solution
We first plot the voltage vs current curve, which looks like a straight line.

Fig. S3.3

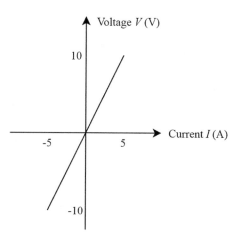

The slope of this straight is $R = 2$ V/A. The measurements consistently follow Ohm's law $V = IR$, with $R = 2\ \Omega$. Hence, this element behaves like a $2\ \Omega$ resistor.

Problem 3.4 True or False?

(a) If you double the voltage across the resistor, the current through it doubles.
(b) If you double the voltage across the resistor, the current through it halves.
(c) If you halve the current through the resistor, the voltage across it doubles.
(d) If you halve the current through the resistor, the voltage across it halves.
(e) If you double the resistance of a resistor and keep the voltage across the resistor unchanged, the current through the resistor doubles.
(f) If you double the resistance of a resistor and keep the current through the resistor unchanged, the voltage across the resistor doubles.

Solution
(a) True.
(b) False. The current doubles.
(c) False. The voltage halves.
(d) True
(e) False. The current halves.
(f) True.

Problem 3.5 The total human body in water is approximately 300 Ω. The electric current over 10 mA is life threatening if the current runs through the heart (10 mA = 0.01 A). How much voltage in the water can be lethal?

Solution
According to Ohm's law,

$$V = IR = (10 \text{ mA})(300 \text{ } \Omega) = 3000 \text{ mV} = 3 \text{ V}.$$

This lethal voltage of 3 V is very low.

The good news is that dry skin has resistance as high as 100 kΩ (1 kΩ = 1000 Ω). Dry and unbroken skin can protect us at low voltage. But high voltage (above 500 V) can breakdown our skin.

Chapter 4. Kirchhoff's Voltage Law (KVL)

Problem 4.1 Using the given voltage polarities, set up KVL equations for the following circuits:

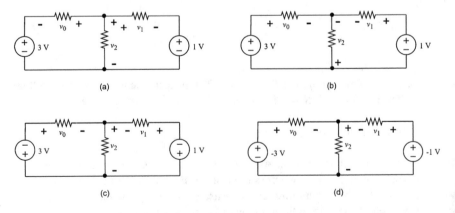

Fig. P4.1

Solution

There are two meshes in each circuit, and we can set up two KVL equations for each case.

(a)
$$-v_0 + v_2 = 3 \text{ V},$$
$$-v_1 + v_2 = 1 \text{ V}.$$

(b)
$$v_0 - v_2 = 3 \text{ V},$$
$$v_1 - v_2 = 1 \text{ V}.$$

(c)
$$-v_0 - v_2 = 3 \text{ V},$$
$$-v_1 - v_2 = 1 \text{ V}.$$

(d)
$$v_0 + v_2 = -3 \text{ V},$$
$$v_1 + v_2 = -1 \text{ V}.$$

Problem 4.2 In this problem, we will use a new current source called controlled current source, whose symbol is a diamond with an arrow inside (see the figure below). The symbol for a regular current source is a circle with an arrow inside. For example, a controlled current source is

Fig. P4.2a

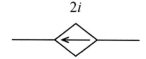
$2i$

Here "$2i$" indicate the value of this current source, and this value is two times the current value i, which is defined elsewhere in the circuit.

Fig. P4.2b

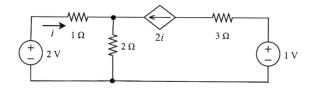

Find the current i in this circuit.

Solution

Since the elements in series with a current source have no contribution to the circuit, we can ignore the 3 Ω resistor and the 1 V voltage source. [Likewise, we can ignore the elements in parallel with a voltage source.]

The current running down the 2 Ω resistor is $3i$, which is the summation of i from the left branch and $2i$ from the right branch.

Using Ohm's law, the voltage across the 1 W resistor is $(1\ \Omega)(i)$.

The voltage across the 2 W resistor is $(2\ \Omega)(3i)$.

The KVL equation for the left mesh is given as

$$(1\ \Omega)(i) + (2\ \Omega)(3i) = 2\text{ V}.$$

That is,

$$7i = 2\text{ V}.$$

Finally,

$$i = 2/7\text{ A}.$$

Problem 4.3 Set up the KVL equations for the following Wheatstone bridge circuit.

Fig. P4.3

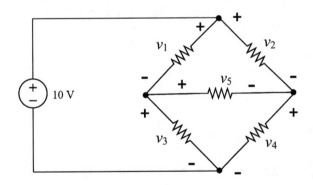

Solution

This circuit consists of three meshes, and we can set up three KVL equations.

$$v_1 + v_3 = 10\text{ V},$$

$$v_1 + v_5 - v_2 = 0,$$

$$v_5 + v_4 - v_3 = 0.$$

Problem 4.4 Do not simplify the circuit. Use the KVL to solve for the current i in the circuit.

Fig. P4.4

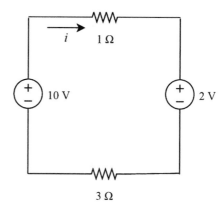

Solution
Using Ohm's law, the voltage drop across the 1 Ω resistor is $(1\ \Omega)(i)$.
  The voltage drop across the 3 Ω resistor is $(3\ \Omega)(i)$.
  The circuit has only one mesh, and we can set up one KVL equation as follows:

$$10\ V = (1\ \Omega)(i) + (2\ V) + (3\ \Omega)(i).$$

After simplification, we obtain

$$8\ V = (4\ \Omega)(i),$$

and

$$i = 2\ V/\Omega = 2\ A.$$

Problem 4.5 Use KVL to verify if the following circuit is valid.

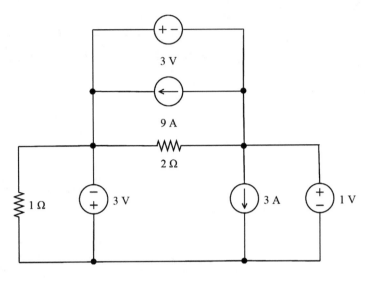

Fig. P4.5

Solution
If an element is connected to a voltage source in parallel, this element does not affect
the value of the voltage source and the voltage across this element is determined by
the voltage source. In order to check whether the circuit is valid, we first simplify the
circuit by removing the elements in parallel with a voltage source, obtaining a
simplified circuit shown in Fig. P5.4.

Fig. S4.5

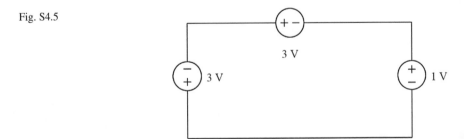

A valid circuit should obey the KVL, namely

$$(3 \text{ V}) + (1 \text{ V}) + (3 \text{ V}) = 0,$$

which obviously is an incorrect equation. Thus, this circuit is invalid.

Problem 4.6 Use KVL to verify if the following circuit is valid.

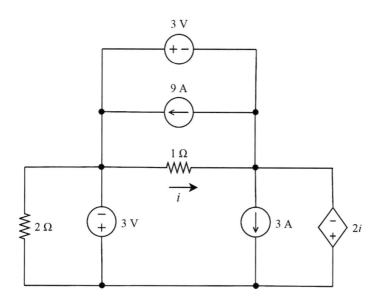

Fig. P4.6

Solution
This circuit contains a controlled voltage source labeled by "$2i$," where "i" is the voltage drop across the 1 Ω resistor. This controlled source is referred to as a *current-controlled voltage source*. The unit of "i" is amperes (A) and the unit of "$2i$" is volts (V). Clearly, the conversion factor "2" has a unit of "volts per ampere" (V/A).

Voltage across the 1 Ω resistor is 3 V provided by the 3 V voltage source. Using Ohm's law, i can be obtained by the product (3 V)/(1 Ω) = (3 A).
The current-controlled voltage source then has a value of $2i = \left(2\frac{V}{A}\right) \times (3 \text{ A}) = 6$ V.
The KVL equation can be readily obtained as

$$(3 \text{ V}) + (3 \text{ V}) = 6 \text{ V},$$

which is valid. Therefore, this is a valid circuit. When we set up this KVL equation, we only pay attention to the voltage sources because the elements connected to the voltage sources in parallel do not affect the values of the voltage sources. We can ignore other elements in parallel with the voltage sources when setting up equations using Fig. S4.6, where the controlled voltage source is replaced by a regular voltage source of 6 V after we figure out the value of the controlled source to be 6 V.

Fig. S4.6

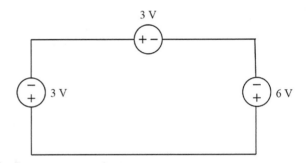

Chapter 5. Kirchhoff's Current Law (KCL)

Problem 5.1 Use KCL to verify if the following circuit is valid.

Fig. P5.1

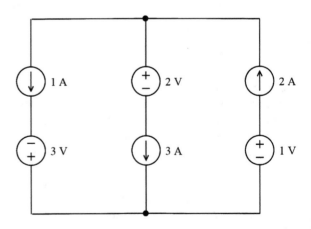

Solution
Let us simplify the circuit by removing all voltage sources connected in series with a current source because if an element is connected in series with a current source, this element can be ignored. Figure P5.1 is then simplified as Fig. S5.1.

Fig. S5.1

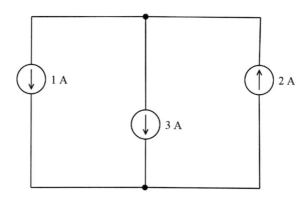

There are two nodes in the circuit, and we can set up a KCL equation at one of these nodes. The total current running into the node should equal to the total current running out from the node. This leads to

$$(1 \text{ A}) + (3 \text{ A}) = 2 \text{ A},$$

which is clearly not valid. Therefore, the circuit is invalid.

Problem 5.2 Use KCL to verify if the following circuit is valid.

Fig. P5.2

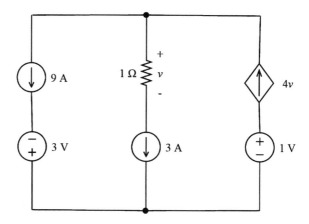

Solution

This circuit contains a controlled current source labeled by "$4v$," where "v" is the voltage drop across the 1 Ω resistor. This controlled source is referred to as a *voltage-controlled current source*. The unit of "v" is volts (V) and the unit of "$4v$" is amperes (A). Clearly, the conversion factor "4" has a unit of "amperes per volt" (A/V).

The current through the 1 Ω resistor is 3 A provided by the 3 A current source. Using Ohm's law, v can be obtained by the product (1 Ω) (3 A) = 3 V.

The voltage-controlled current source then has a value of $4v = 4$ (A/V) $\times$ (3 V) = 12 A.

The KCL equation can be readily obtained as

$$(9 \text{ A}) + (3 \text{ A}) = 12 \text{ A},$$

which is valid. Therefore, this is a valid circuit. When we set up this KCL equation, we only pay attention to the current sources because the elements connected to the current sources in series do not affect the values of the current sources. If a circuit branch contains a current source, the current in that branch is determined only by the current source, and we can ignore other elements in that branch when setting up equations.

Problem 5.3 Set up the KCL equations for the following Wheatstone bridge circuit.

Fig. P5.3

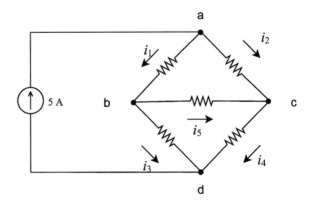

Solution

This circuit has four nodes, and we can set up three KCL equations at nodes a, b, and c.

$$a : i_1 + i_2 = 5 \text{ A},$$

$$b : i_1 = i_3 + i_5,$$

$$c : i_2 + i_5 = i_4.$$

We do not need to set up a KCL equation at node d, because this equation does not provide any new information about the circuit. In fact, if we substitute the equation at node b and the equation at node c into the equation at node a, we get

$$i_3 + i_4 = 5 \text{ A},$$

which is the KCL equation at node d.

In general, if the circuit has n nodes, we can only get n-1 independent KCL equations.

Problem 5.4 Find the current i_1 in the circuit shown in Fig. P5.4.

Fig. P5.4

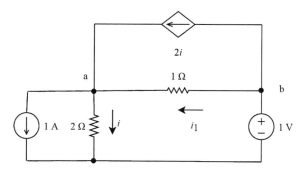

Solution
This circuit has three nodes, and we would normally set up two KCL equations. However, if a node is directly connected to a voltage source, we do not set up a KCL equation for that node. Likewise, if a loop contains a current source, we do not set up a KVL equation for that loop.

We only need to set up one KCL equation at node a as

$$(2i) + i_1 = i + (1 \text{ A}),$$

that is,

$$i + i_1 = 1 \text{ A.}$$

In this equation, we have two unknowns i and i_1. We do not have enough equations to solve for the unknown i_1.

There are three meshes in the circuit, and we normally set up two KVL equations for this circuit. However, two of the three meshes contains a current source. One mesh contains an independent 1 A current source and the other mesh contains a controlled (dependent) "$2i$" current source. We can only set up one KVL in the loop that contains a 1 V voltage source and two resistors.

Using Ohm's law, the voltage drop across the 1 Ω resistor is $(1 \; \Omega) (i_1)$, and the voltage drop across the 2 Ω resistor is $(2 \; \Omega) (i)$. Thus, the KVL equation is

$$(1 \; \Omega) \times i_1 + (2 \; \Omega) \times i = 1 \text{ V.}$$

We now have a system of two equations with two unknowns i and i_1.

$$\begin{cases} i + i_1 = 1 \text{ A} \\ (1 \; \Omega) \times i_1 + (2 \; \Omega) \times i = 1 \text{ V.} \end{cases}$$

Divided by $(1 \; \Omega)$ on both sides of the second equation, the system becomes

$$\begin{cases} i + i_1 = 1 \text{ A} \\ 2i + i_1 = 1 \text{ A.} \end{cases}$$

This system has a solution of

$$\begin{cases} i = 0 \text{ A} \\ i_1 = 1 \text{ A.} \end{cases}$$

Problem 5.5 This circuit model a transistor, which has many applications such as amplifiers. Find i_b.

Fig. P5.5

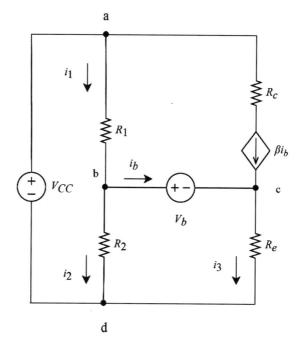

Solution

We need to use both KVL and KCL for this problem.

There are three meshes in the circuit, and we normal can set up three KVL equations. The upper-left mesh contains a controlled current source, and we do not set up a KVL equation for that mesh.

There are four nodes in the circuit, and we set up three KCL. We generally want to avoid KCL equations that involve voltage sources. However, i_b is what we are looking for and i_b is used in the controlled source calculation. This problem forces us to set up equations that include i_b. Therefore, the KCL equations at nodes b and c are set up. We thus have a system of four equations:

$$\begin{cases} \text{KVL} : R_1 i_1 + R_2 i_2 = V_{cc} \\ \text{KVL} : R_2 i_2 = V_b + R_e i_3 \\ \text{KCL at b} : i_1 = i_2 + i_b \\ \text{KCL at c} : i_b + \beta i_b = i_3 \end{cases}$$

Next, we substitute the KCL equations into KVL equations and obtain

$$\begin{cases} \text{KVL} : R_1(i_2 + i_b) + R_2 i_2 = V_{cc} \\ \text{KVL} : R_2 i_2 = V_b + R_e(1 + \beta)i_b \end{cases}$$

Rearranging the terms yields

$$\begin{cases} (R_1 + R_2)i_2 = V_{cc} - R_1 i_b \\ R_2 i_2 = V_b + R_e(1 + \beta)i_b \end{cases}$$

Eliminating i_2 yields

$$\frac{V_{cc} - R_1 i_b}{R_1 + R_2} = \frac{V_b + R_e(1 + \beta)i_b}{R_2}$$

Finally,

$$i_b = \frac{\frac{V_{cc}}{R_1 + R_2} - \frac{V_b}{R_2}}{\frac{R_e(1+\beta)}{R_2} + \frac{R_1}{R_1 + R_2}}.$$

Chapter 6. Resistors in Series and in Parallel

Problem 6.1 Ten 1 kΩ resistors are connected in series, the total resistance is

(a) 10 Ω
(b) 100 Ω
(c) 1 kΩ
(d) 10 kΩ
(e) 100 kΩ

Solution
(d) 10 kΩ
  For ten 1 kΩ resistors in series, the total resistance is

$$R_{\text{total}} = 10 \times (1 \text{ k}\Omega) = 10 \text{ k}\Omega.$$

Problem 6.2 Ten 1 kΩ resistors are connected in parallel, the total resistance is

(a) 10 Ω
(b) 100 Ω
(c) 1 kΩ
(d) 10 kΩ
(e) 100 kΩ

Solution
(b) $100\ \Omega$
For ten $1\ k\Omega$ resistors in parallel, the total resistance is

$$\frac{1}{R_{total}} = 10 \times \frac{1}{1\ k\Omega} = \frac{1}{100\ \Omega}.$$

Thus,

$$R_{total} = 100\ \Omega.$$

Problem 6.3 Two resistors R_1 and R_2 are connected in series. The total resistance is $1\ k\Omega$.

(a) R_1 is less than $1\ k\Omega$.
(b) R_1 is larger than $1\ k\Omega$.
(c) R_1 is $500\ \Omega$.
(d) R_1 and R_2 must have the same resistance.
(e) R_1 and R_2 must not have the same resistance.

Solution
(a) R_1 is less than $1\ k\Omega$.
When R_1 and R_2 are connected in series, the total resistance is

$$R_{total} = R_1 + R_2 = 1\ k\Omega.$$

Since R_1 and R_2 are positive, none of them can be larger than $1\ k\Omega$.

Problem 6.4 Two resistors R_1 and R_2 are connected in parallel. The total resistance is $1\ k\Omega$.

(a) R_1 is less than $1\ k\Omega$.
(b) R_1 is larger than $1\ k\Omega$.
(c) R_1 is $2\ k\Omega$.
(d) R_1 and R_2 must have the same resistance.
(e) R_1 and R_2 must not have the same resistance.

Solution
(b) R_1 is larger than $1\ k\Omega$.
When resistors are connected in parallel, the total resistance is smaller than the smallest resistor. If the total resistance is $1\ k\Omega$, then both R_1 and R_2 must be larger than $1\ k\Omega$.

Problem 6.5 Four resistors R_1, R_2, R_3, and R_4 are connected in parallel. They satisfy the relationship: $R_1 = R_2 < R_3 = R_4$. The total resistance is $1\ k\Omega$.

(a) R_1 is less than 1 kΩ.
(b) R_1 is less than 2 kΩ.
(c) R_3 is less than 1 kΩ.
(d) R_3 is less than 2 kΩ.
(e) R_3 is less than 4 kΩ.
(f) R_1 is larger than 4 kΩ.
(g) None of the above.

Solution
(e) None of the above.

The total resistance is calculated as

$$\frac{1}{R_{\text{total}}} = \frac{1}{R_1} + \frac{1}{R_2} + \frac{1}{R_3} + \frac{1}{R_4} = \frac{2}{R_1} + \frac{2}{R_3}.$$

Let us consider the constraints for R_1.
It is given that $R_{total} = 1$ kΩ and $R_1 < R_3$, and then

$$\frac{1}{1k\Omega} = \frac{2}{R_1} + \frac{2}{R_3} < \frac{2}{R_1} + \frac{2}{R_1} = \frac{4}{R_1}.$$

That is,

$$\frac{1}{4\ \text{k}\Omega} < \frac{1}{R_1}.$$

$$R_1 < 4\ \text{k}\Omega.$$

On the other hand, $R_3 > 0$, then

$$\frac{1}{1\ \text{k}\Omega} = \frac{2}{R_1} + \frac{2}{R_3} > \frac{2}{R_1}.$$

That is,

$$\frac{1}{2\ \text{k}\Omega} > \frac{1}{R_1}.$$

$$R_1 > 2\ \text{k}\Omega.$$

Finally,

$$2\ \text{k}\Omega < R_1 < 4\ \text{k}\Omega.$$

Let us consider the constraints for R_3.
Since $R_1 < R_3$,

$$\frac{1}{1k\Omega} = \frac{2}{R_1} + \frac{2}{R_3} > \frac{2}{R_3} + \frac{2}{R_3} = \frac{4}{R_3}.$$

That is,

$$\frac{1}{4\ k\Omega} > \frac{1}{R_3}.$$

$$R_3 > 4\ k\Omega.$$

Problem 6.6 Ten resistors are connected in parallel, with $R_n = n\ \Omega$, for $n = 1, 2, \ldots,$ 10.

(a) $R_{\text{total}} = 1 + 2 + 3 + 4 + 5 + 6 + 7 + 8 + 9 + 10 = 55\ \Omega$

(b) $R_{\text{total}} = \left(\frac{1}{1} + \frac{1}{2} + \frac{1}{3} + \frac{1}{4} + \frac{1}{5} + \frac{1}{6} + \frac{1}{7} + \frac{1}{8} + \frac{1}{9} + \frac{1}{10}\right) = 2.9290\ \Omega$

(c) $R_{\text{total}} < 1\ \Omega$

(d) $R_{\text{total}} = \frac{1\times2\times3\times4\times5\times6\times7\times8\times9\times10}{1+2+3+4+5+6+7+8+9+10}\ \Omega$

Solution

(c) $R_{\text{total}} < 1\ \Omega$

The total resistance is calculated as

$$\frac{1}{R_{\text{total}}} = \frac{1}{R_1} + \frac{1}{R_2} + \frac{1}{R_3} + \cdots + \frac{1}{R_{10}} = \frac{1}{1} + \frac{1}{2} + \frac{1}{3} + \cdots + \frac{1}{10}.$$

$$R_{\text{total}} = \frac{1}{\frac{1}{R_1} + \frac{1}{R_2} + \frac{1}{R_3} + \cdots + \frac{1}{R_{10}}} = \frac{1}{\frac{1}{1} + \frac{1}{2} + \frac{1}{3} + \cdots + \frac{1}{10}}.$$

$$R_{\text{total}} < \frac{1}{\frac{1}{10} + \frac{1}{10} + \frac{1}{10} + \cdots + \frac{1}{10}} = 1\ \Omega.$$

The total resistance is always smaller than the smallest resistor when the resistors are connected in parallel.

Problem 6.7 Find the total resistance for the resistor network shown in Fig. P6.7. Each resistor in the network is $1\ \Omega$.

Fig. P6.7

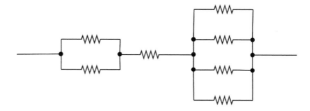

Solution
This resistor network can be divided into three sections. The right section contains four resistors connected in parallel as shown in Fig. S6.7a. Since all four resistors are the same, and each has a resistance of 1 Ω. The equivalent resistance is 1/4 Ω.

Fig. S6.7a

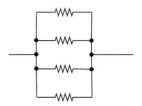

Similarly, the left section contains two resistors connected in parallel as shown in Fig. S6.7b. These two resistors have the same value of 1 Ω. Thus, the equivalent resistance of these two resistors is 1/2 Ω.

Fig. S6.7b

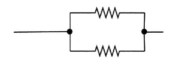

Finally, these three sections can be represented as three resistors connected in series as shown in Fig. S6.7c. The total resistance is the sum

$$R_{total} = (0.5\ \Omega) + (1\ \Omega) + (0.25\ \Omega) = 1.75\ \Omega.$$

$$
\begin{array}{ccc}
0.5\ \Omega & 1\ \Omega & 0.25\ \Omega \\
-\text{\Large WW} & -\text{\Large WW} & -\text{\Large WW}-
\end{array}
$$

Fig. S6.7c

Chapter 7. Voltage Divider and Current Divider

Problem 7.1 In the circuit shown in Fig. P7.1, $R_1 = 1$ kΩ. Find the values of other resistors.

Fig. P7.1

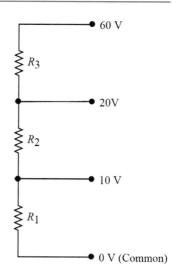

Solution

This is a voltage divider circuit.

Resistors R_1 and R_2 divide the voltage in half. In other words, the voltage drop on R_1 is the same as the voltage drop on R_2. This implies that $R_2 = R_1 = 1$ kΩ.

The voltage drop on R_3 is 40 V, which is four times the voltage drop on R_1. Therefore,

$$R_4 = 4R_1 = 4 \times 1 \text{ k}\Omega = 4 \text{ k}\Omega.$$

We must point out that the voltage divider technique can only be applied when the resistors involved have the same current. Do not apply the voltage divider technique if the resistors have different currents. In our problem, the voltage terminals labeled "10 V," "20 V," and "60 V" do not connect to other elements. If other elements are connected to the voltage terminals, the voltage divider technique cannot be used.

Problem 7.2 Find the voltage v in the circuit shown in Fig. P7.2. All resistors have the value of 1 Ω.

Fig. P7.2

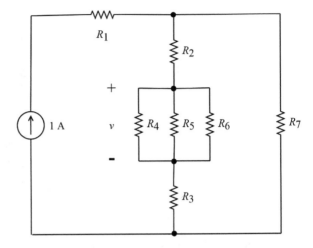

Solution
First of all, R_1 is in series of the current source, and it does not affect the current of the source. Therefore, R_1 can be removed without affecting the rest of the circuit.

Resistors R_4, R_5, and R_6 are in parallel, and each of them has a value of 1 Ω. The combined resistor is labeled as R_{456}. The value of these three resistors is 1/3 Ω.
Resistors R_2 and R_3 are in series, and each of them has a value of 1 Ω. The combined value of these two resistors is labeled as R_{23}. The value of R_{23} is 2 Ω.
Figure P7.2 can be reduced to Fig. S7.2a.

Fig. S7.2a

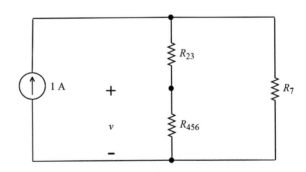

Figure S7.2 is further reduced to Fig. S7.2b, by combining resistors R_{23} and R_{456}. We denote the combed resistor as R_{23456}. Since $R_{23} = 2\ \Omega$ and $R_{456} = 1/3\ \Omega$, $R_{23456} = R_{23} + R_{456} = 2 + 1/3 = 7/3\ \Omega$.

Fig. S7.2b

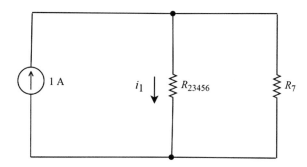

The circuit in Fig. S7.2b is a typical current divider, the current i_1 can be calculated as

$$i_1 = (1\ \text{A})\frac{R_7}{R_7 + R_{23456}} = (1\ \text{A})\frac{(1\ \Omega)}{(1\Omega) + (7/3\ \Omega)} = \frac{3}{10}\ \text{A}.$$

Finally, using Ohm's law, we have (see Fig. S7.2a)

$$v = i_1 \times R_{456} = \left(\frac{3}{10}\ \text{A}\right) \times \left(\frac{1}{3}\ \Omega\right) = \frac{1}{10}\ \text{V} = 0.1\ \text{V}.$$

Problem 7.3 Calculate currents i_1, i_2, i_3, and i_4 in the circuit shown in Fig. P7.3.

(a) $R_1 = R_2 = R_3 = R_4 = 1\ \Omega$.
(b) $R_1 = 1\ \Omega$, $R_2 = 2\ \Omega$, $R_3 = 3\ \Omega$, and $R_4 = 4\ \Omega$.

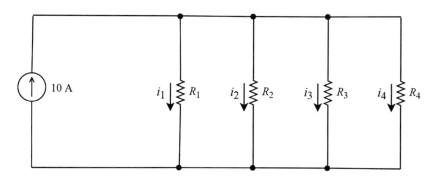

Fig. P7.3

Solution

(a) The condition $R_1 = R_2 = R_3 = R_4 = 1\,\Omega$ leads to

$$i_1 = i_2 = i_3 = i_4 = \frac{10\,\text{A}}{4} = 2.5\,\text{A}.$$

If all the resistors are of the same value, they share the same amount of current.

(b) We need to use the current divider formula four times for this part.

$$i_1 = (10\,\text{A})\frac{\frac{1}{R_1}}{\frac{1}{R_1}+\frac{1}{R_2}+\frac{1}{R_3}+\frac{1}{R_4}} = (10\,\text{A})\frac{\frac{1}{R_1}}{\frac{1}{1}+\frac{1}{2}+\frac{1}{3}+\frac{1}{4}} = \frac{24}{5}\times\frac{1}{R_1} = \frac{24}{5}\,\text{A}$$

$$i_2 = (10\,\text{A})\frac{\frac{1}{R_2}}{\frac{1}{R_1}+\frac{1}{R_2}+\frac{1}{R_3}+\frac{1}{R_4}} = (10\,\text{A})\frac{\frac{1}{R_2}}{\frac{1}{1}+\frac{1}{2}+\frac{1}{3}+\frac{1}{4}} = \frac{24}{5}\times\frac{1}{R_2} = \frac{12}{5}\,\text{A}$$

$$i_3 = (10\,\text{A})\frac{\frac{1}{R_3}}{\frac{1}{R_1}+\frac{1}{R_2}+\frac{1}{R_3}+\frac{1}{R_4}} = (10\,\text{A})\frac{\frac{1}{R_3}}{\frac{1}{1}+\frac{1}{2}+\frac{1}{3}+\frac{1}{4}} = \frac{24}{5}\times\frac{1}{R_3} = \frac{8}{5}\,\text{A}$$

$$i_4 = (10\,\text{A})\frac{\frac{1}{R_4}}{\frac{1}{R_1}+\frac{1}{R_2}+\frac{1}{R_3}+\frac{1}{R_4}} = (10\,\text{A})\frac{\frac{1}{R_4}}{\frac{1}{1}+\frac{1}{2}+\frac{1}{3}+\frac{1}{4}} = \frac{24}{5}\times\frac{1}{R_4} = \frac{6}{5}\,\text{A}$$

Problem 7.4 Find i_1 and i_2 in the circuit shown in Fig. P7.4.

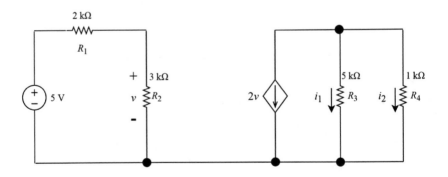

Fig. P7.4

Solution

The left section of the circuit is a voltage divider. The resistor gets its voltage proportional to it resistance value. The ratio of $R_2{:}R_1$ is 3:2. The voltage source distributes 5 V to R_2 and R_1 according to this ratio. As a result, R_2 gets 3 volts, and R_1 gets 2 volts. Therefore,

$$v = 3 \text{ V}.$$

The right section of the circuit is a current divider. The voltage-controlled current source generates a current of $2v = 6$ A.

For a current divider, the total current is distributed to the resistors according to the ratio of conductance. The conductance is the reciprocal of the resistance. The total current of 6A is distributed to R_3 and R_4 according to the ratio

$$\frac{1}{R_3} : \frac{1}{R_4} = \frac{1}{5} : \frac{1}{1} = 1 : 5.$$

In other words, R_4 gets five times more current than R_3 gets. Therefore,

$$i_1 = 1 \text{ A,}$$

$$i_2 = 5 \text{ A.}$$

For a voltage divider, a larger resistor gets a larger portion of the voltage.
For a current divider, a larger resistor gets a smaller portion of the current.
You may notice that there is a wire connecting the left section and the right section in this circuit. There is no current in this wire because current can only flow in a complete loop, and this wire is not part of a loop.

Problem 7.5 For a voltage divider circuit shown in Fig. P7.5. R_1 is 1 Ω. The voltage across R_1 is 1 V. Is it possible to determine the source voltage v and the value of the other resistor R_2?

Fig. P7.5

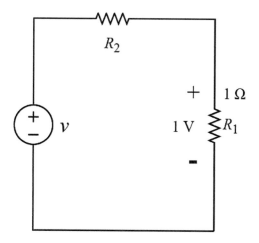

Solution

Using the voltage divider formula, we have

$$1 \text{ V} = v \, \frac{R_1}{R_1 + R_2} = v \, \frac{(1\Omega)}{(1\Omega) + R_2}.$$

We only one equation but two unknowns v and R_2. We are unable to obtain a unique solution. In fact, we can have infinite number of solutions. Here are three solutions:

$$R_2 = 1 \, \Omega, v = 2 \text{ V};$$

$$R_2 = 2 \, \Omega, v = 3 \text{ V};$$

$$R_2 = 3 \, \Omega, v = 4 \text{ V}.$$

Chapter 8. Node-Voltage Method

Problem 8.1 Set up node equations for the circuit given in Fig. P8.1.

Solution

The ground is selected as the reference node. We will calculate the voltages at the other two nodes, noted as v_1 and v_2. The two node equations are set up as follows.

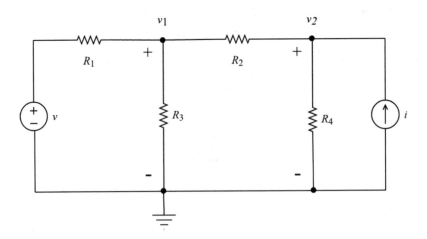

Fig. P8.1

$$\frac{v_1 - v}{R_1} + \frac{v_1}{R_3} + \frac{v_1 - v_2}{R_2} = 0,$$

$$\frac{v_2 - v_1}{R_2} + \frac{v_2}{R_4} = i.$$

Here, R_1, R_2, R_3, R_4, v, and i are given. The two unknowns are v_1 and v_2. The node equations are essentially KCL equations. Each term in the node equations is a current.

Problem 8.2 Set up node equations for a circuit containing a controlled source.

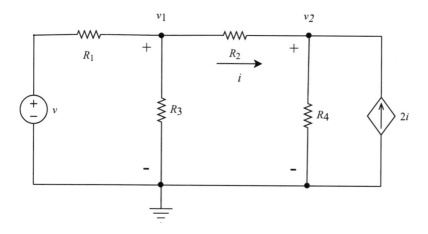

Fig. P8.2

Solution

The procedure is almost the same as in Problem 8.1, except that one extra equation is needed for the relationship regarding the controlled source.

The ground is selected as the reference node. We will calculate the voltages at the other two nodes, noted as v_1 and v_2. The two node equations are set up as follows.

$$\frac{v_1 - v}{R_1} + \frac{v_1}{R_3} + \frac{v_1 - v_2}{R_2} = 0.$$

$$\frac{v_2 - v_1}{R_2} + \frac{v_2}{R_4} = 2i.$$

We need one more equation for the controlled source relationship:

$$i = \frac{v_1 - v_2}{R_2}.$$

Here, R_1, R_2, R_3, R_4, and v are given. The unknowns are v_1 and v_2.

Problem 8.3 Set up the node equations for a circuit, in which a voltage source is between the two nodes, and there are no resistors between these two nodes.

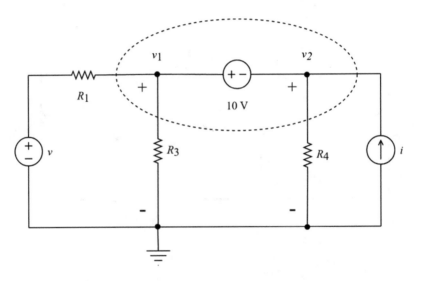

Fig. P8.3

Solution
If there is nothing but a voltage source between two nodes, the regular node equations cannot be established because we do not know how to express the current between these two nodes. A well accepted remedy is to use a super node (as indicated by a dotted ellipse in Fig. P8.3).

We will only have one unknown, v_2. The node voltage v_1 will be expressed as $v_2 + 10$ V. The following is the node equation for the super node:

$$\frac{v_2 + 10 - v}{R_1} + \frac{v_2 + 10}{R_3} + \frac{v_2}{R_4} = i.$$

Problem 8.4 Set up the node equations for a circuit, in which a controlled voltage source is between the two nodes, and there are no resistors between these two nodes.

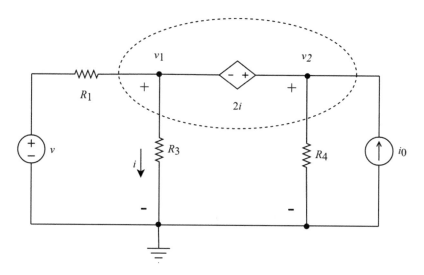

Fig. P8.4

Solution

If there is nothing but a controlled voltage source between two nodes, we will use a super node (as indicated by a dotted ellipse in Fig. P8.4). The controlled source needs an extra equation to describe the controlling information. We need to define "i" to calculate "$2i$."

We will only have one unknown, v_1. The node voltage v_2 will be expressed as $v_2 + 2i$. The following is the node equation for the super node:

$$\frac{v_1 - v}{R_1} + \frac{v_1}{R_3} + \frac{v_1 + 2i}{R_4} = i_0.$$

We need one more equation for the controlled source relationship:

$$i = \frac{v_1}{R_3}.$$

There is not much difference in setting up equation for independent sources and for controlled sources. When a controlled source is used, an extra equation is required to describe the depending variable.

Problem 8.5 Set up node equations for the circuit, where a voltage source is between a node and the reference node.

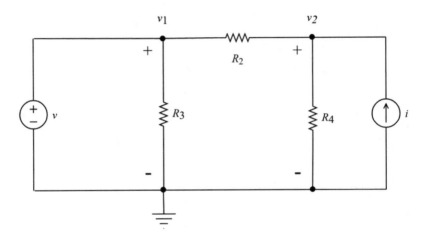

Fig. P8.5

Solution
In this circuit, a voltage source is between a node and the reference node. In this case, the voltage of v_1 is known, and there is no need to set up a node equation for v_1. We only need one node equation for v_2, which is the only unknown to be solved.

$$\frac{v_2 - v}{R_2} + \frac{v_2}{R_4} = i.$$

Chapter 9. Mesh-Current Method

Problem 9.1 Set up the mesh equations and solve for the mesh currents.

Fig. P9.1

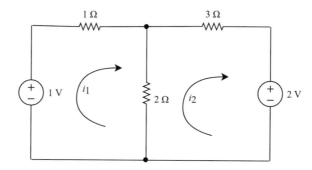

Solution

Two meshes have two mesh equations. Let the mesh current for the left mesh be i_1 and the mesh current for the right mesh be i_2. We have

$$(1\,\Omega)(i_1) + (2\,\Omega)(i_1 - i_2) = 1\text{ V},$$

$$(3\,\Omega)(i_2) + (2\,\Omega)(i_2 - i_1) = -2\text{ V}.$$

Rearranging the terms yields

$$3i_1 - 2i_2 = 1,$$

$$-2i_1 + 5i_2 = -2.$$

Let us solve this system of equations by hand. Multiplying the first equation by 2 and multiplying the second equation by 3, the above two equations become

$$6i_1 - 4i_2 = 2,$$

$$-6i_1 + 15i_2 = -6.$$

Summing these two equations leads to

$$11i_2 = -4.$$

$$i_2 = -\frac{4}{11}\text{ A}.$$

Current i_1 can be solved from $3i_1 - 2i_2 = 1$,

$$3i_1 - 2\left(-\frac{4}{11}\right) = 1,$$

$$3i_1 = 1 - \frac{8}{11} = \frac{3}{11},$$

$$i_1 = \frac{13}{11} \text{ A.}$$

Problem 9.2 This circuit contains a voltage-controlled voltage source. Set up the mesh equations and solve for the mesh currents.

Fig. P9.2

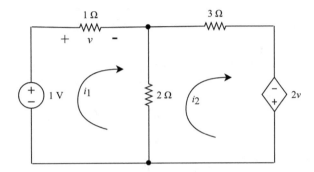

Solution
We can treat the controlled source as a regular source and set up the mesh equations.

Two meshes have two mesh equations. Let the mesh current for the left mesh be i_1 and the mesh current for the right mesh be i_2. We have

$$(1\ \Omega)(i_1) + (2\ \Omega)(i_1 - i_2) = 1 \text{ V},$$

$$(3\ \Omega)(i_2) + (2\ \Omega)(i_2 - i_1) = 2v.$$

The controlled source needs the value v, which can be calculated by

$$v = (1\ \Omega)(i_1),$$

$$2v = 2i_1.$$

Then the two mesh equations can be rewritten as

$$(1\ \Omega)(i_1) + (2\ \Omega)(i_1 - i_2) = 1 \text{ V},$$

$$(3\ \Omega)(i_2) + (2\ \Omega)(i_2 - i_1) = 2i_1.$$

Rearranging the terms yields

$$3i_1 - 2i_2 = 1,$$

$$-4i_1 + 5i_2 = 0.$$

Let us solve this system of equations by hand. Multiplying the first equation by 4 and multiplying the second equation by 3, the above two equations become

$$12i_1 - 8i_2 = 4,$$

$$-12i_1 + 15i_2 = 0.$$

Summing these two equations leads to

$$7i_2 = 4.$$

$$i_2 = \frac{4}{7} \; A.$$

Current i_1 can be solved from $-4i_1 + 5i_2 = 0$.

$$-4i_1 + 5\left(\frac{4}{7}\right) = 0,$$

$$4i_1 = \frac{20}{7},$$

$$i_1 = \frac{5}{7} \; A.$$

Problem 9.3 This circuit contains a current source. Set up the mesh equations and solve for the mesh currents.

Fig. P9.3

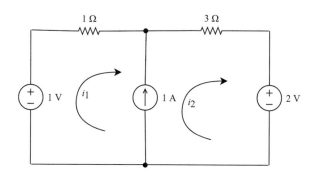

Solution

Since we do not know how to express the voltage drop across a current source, we want to avoid current source in our mesh equations. Using a super mesh can avoid the current sources. For this problem, a super mesh is indicated in Fig. S9.3 as a dotted loop.

Fig. S9.3

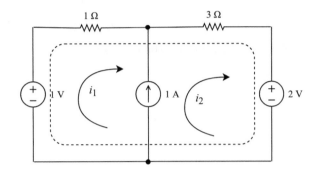

In Fig. S9.3, i_1 and i_2 are related as

$$i_2 - i_1 = 1 \text{ A}.$$

We have only one mesh equation for this super mesh:

$$(1\Omega)(i_1) + (3\Omega)(i_2) = 1 \text{ V} - 2 \text{ V}.$$

Combining the two above equations yields

$$i_1 + 3(1 + i_1) = -1,$$

$$4i_1 = -4,$$

$$i_1 = -1 \text{ A}.$$

Using $i_2 - i_1 = 1$ A, we have

$$i_2 = 0 \text{ A}.$$

Problem 9.4 What if the current source is a controlled current source?

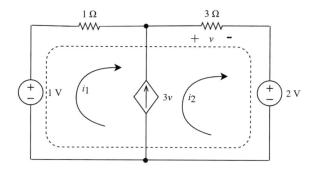

Fig. P9.4

Solution

We use exactly the same strategy to deal with a controlled current source as with an independent current source. Using a super mesh is able to avoid a current source. The dotted loop in Fig. P9.4 indicates a super mesh, which does not contain the controlled current source. The controlling variable is v, which can be calculated via Ohm's law as

$$v = (3\ \Omega)(i_2).$$

In Fig. P9.4, i_1 and i_2 are related as

$$i_2 - i_1 = 3v = 9i_2,$$

$$i_1 = -8i_2.$$

We have only one mesh equation for this super mesh:

$$(1\ \Omega)(i_1) + (3\ \Omega)(i_2) = 1\ \text{V} - 2\ \text{V}.$$

Combining the two above equations yields

$$-8i_2 + 3i_2 = -1,$$

$$-5i_2 = -1,$$

$$i_2 = \frac{1}{5}\ \text{A}.$$

Using $i_1 = -8i_2$, we have

$$i_1 = -\frac{8}{5} \text{ A.}$$

Problem 9.5 Application of the mesh equations, considering a special case of a current course in the circuit.

Fig. P9.5

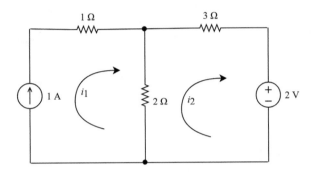

Solution

This current source is at a special location. The mesh current is the same as the current source value. In this case, the mesh current is already determined. There is no need to set up an equation for the mesh current i_1. We already know that

$$i_1 = 1 \text{ A.}$$

We only need one mesh current equation in terms of i_2.

$$(3 \, \Omega)(i_2) + (2 \, \Omega)(i_2 - 1) = -2 \text{ V.}$$

Rearranging the terms yields

$$-2 + 5i_2 = -2,$$

$$5i_2 = -4,$$

$$i_2 = -\frac{4}{5} \text{ A.}$$

Chapter 10. Computer Simulation Software Multisim

Problem 10.1 Use Multisim to simulate a circuit shown to determine the node voltage. You can choose any resistor values.

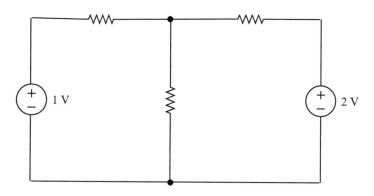

Fig. P10.1

Solution
Run Multisim on a computer. We will create a new project.

File → New → Blank
Place → Component
A new window pops out for you to select components.
To find resistors, you can type "resistor" in the search area or select "Basic" under "Group."

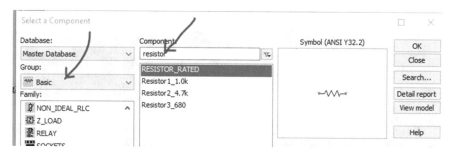

After clicking "OK," a resistor will be placed in the workplace.
To find voltage sources, you can type "DC_POWER" in the search area or select "Sources" under "Group."
After clicking "OK," a dc voltage source will be placed in the workplace.

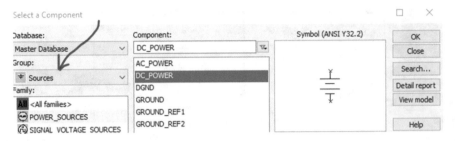

You need three resistors and two voltage sources.

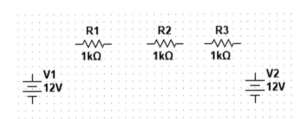

To rotate a component, you right-click and select "Rotate 90°."

To wire up the circuit, go to Place → Wire and connect the components one by one.

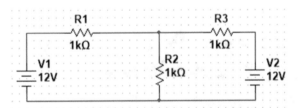

To change the value of a voltage source, double-click or right-click the component and then select "Properties". Change the voltage and click "OK."

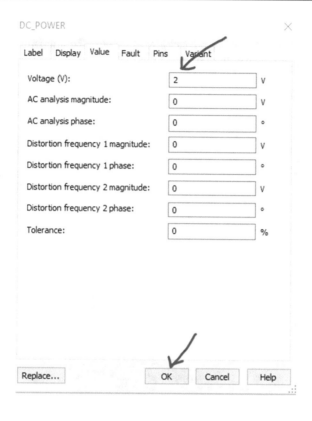

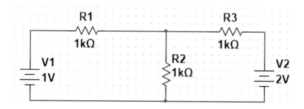

Select a multimeter from the right column. Connect to multimeter to the circuit.

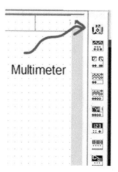

Find a "GROUND" by going to Place → Components. Under "Group," select "Sources." Put two GROUNDs in the circuit as shown. The simulation will not run if the circuit does not have a ground.

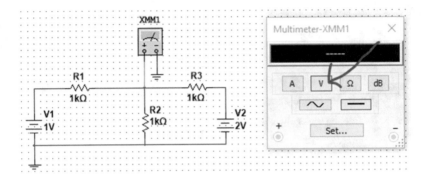

Click the multimeter and configure it to be a voltmeter.

To simulate the circuit and find the node voltage, go to Simulate → Run (or simply click the green triangle on the tool bar). You will see the result of the simulation in the voltmeter.

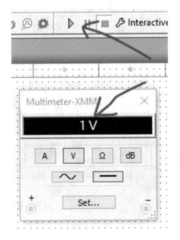

To stop the simulation, go to Simulate → Stop (or simply click the red square on the tool bar).

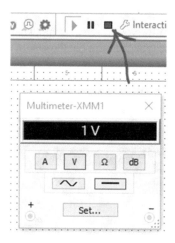

Now, change some resistor values and repeat the experiment.

Problem 10.2 Chapter 15 discusses operational amplifiers (Op-Amps). Simulate an inverting amplifier with a sinewave input. Change the power supply values to observe any potential clipping in the output. The op-amp circuit is given in Fig. P10.2.

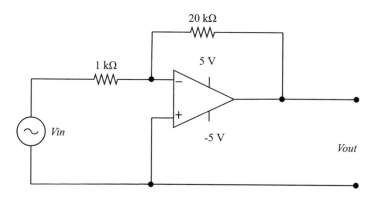

Fig. P10.2

Solution
Run Multisim on a computer. We will create a new project.

File → New → Blank

Place $\rightarrow$ Component

A new window pops out for you to select components.

To find resistors, you can type "resistor" in the search area or select "Basic" under "Group."

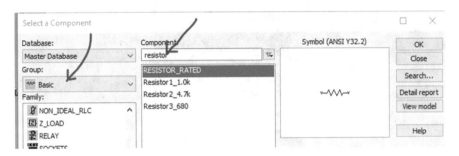

After clicking "OK," a resistor will be placed in the workplace. You need two resistors.

To find voltage sources, you can type "DC_POWER" in the search area or select "Sources" under "Group."

After clicking "OK," a dc voltage source will be placed in the workplace. You need two DC voltage sources. You need to flip one dc voltage source upside down by right-clicking it and rotating it 90° twice.

To find an AC (alternating current) source, you can type "AC_POWER" in the search area or select "Sources" under "Group."

After clicking "OK," an AC voltage source will be placed in the workplace.

An op-amp can be found by searching in Group $\rightarrow$ Analog $\rightarrow$ OPAMP.

There are many options to choose from. You can pick, for example, "741." Click "OK."

Get two GROUNDs under "Sources."

Change the values of the components by clicking the component and inputting the correct values, which are shown in the following schematic.

The op-amp wiring is rather confusing. The co-amp is first flipped upside down. The inverting input $(-)$ "pin 2" is connected to the two resistors. The non-inverting input $(+)$ "pin 3" is connected to the ground. The output "pin 6" is connected to the feedback resistor. A positive power supply is connected to "pin 7." A negative power supply is connected to "pin 4."

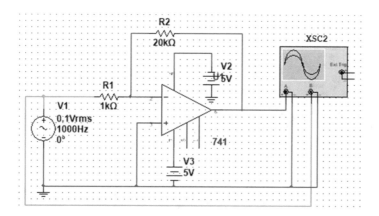

An oscilloscope with two inputs can be found on the right column. "Input A" of the oscilloscope is used to monitor the op-amp output. "Input B" of the oscilloscope is used to monitor the op-amp input.

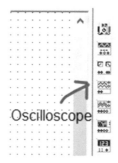

Clicking on the oscilloscope symbol opens the oscilloscope display.

Run a simulation by clicking the green triangle on the tool bar. In order to better visualize the waveforms in the oscilloscope display, you are encouraged to adjust the scaling setting for Timebase, Channel A, and Channel B as indicated in the following figure.

In the oscilloscope display, the red wave is the output and the green wave is the input. The default color is red. To get a green curve, you need to right-click the wire that is attached to the input source. Selecting the "Property" option. Change the "Net color" to green.

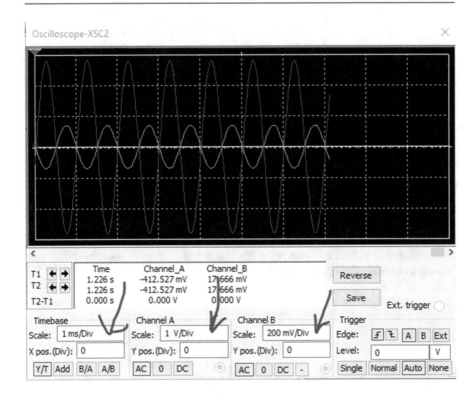

Next, we change the negative power supply to 0 V as indicated below. An equivalent way is to connect the op-amp's "pin4" to the ground. After running another simulation with this change, we observed that the output is clipped.

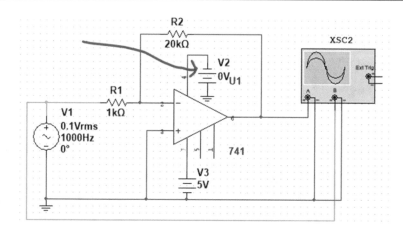

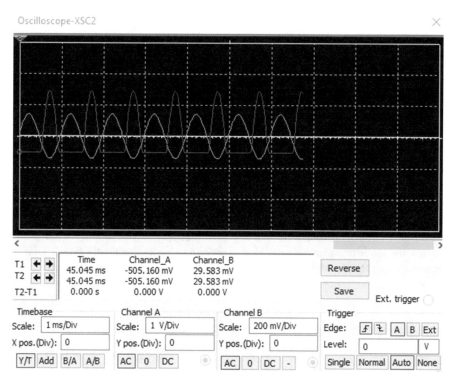

Finally, we change both voltage sources to 1 V and run a simulation. The results are shown below. The output waveform is clipped on both positive and negative directions.

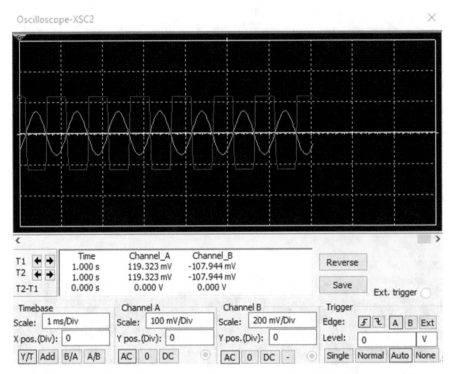

Problem 10.3 Operational amplifier circuits are normally designed to operate from dual supplies, e.g., +9 V and − 9 V. This is not always easy to achieve and therefore it is often convenient to use a single-ended or single supply version of the electronic circuit design. Find a single supply op-amp circuit and use Multisim circuit.

Solution

The following circuit uses only one 9 V power supply. It can amplify a sinewave signal by introducing a positive bias. The output is an amplified signal with a positive dc bias. The Multisim circuit and simulation result are shown below.

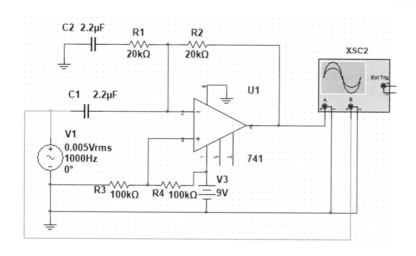

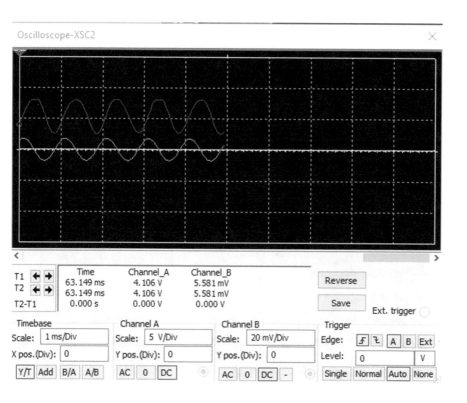

Chapter 11. Superposition

Problem 11.1 A student uses the superposition principle to solve the voltage v. The student's answer is **wrong**. Please help this student to find the mistake. Let us start with Fig. P11.1a.

Fig. P11.1a

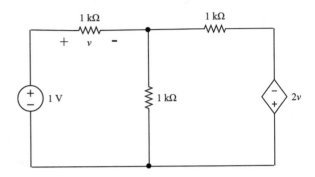

There are two sources in the circuit.

Case 1: Let us first remove the source on the left, obtaining Fig. P11.1b. It can be verified that the solution is $v = 0$.

Fig. P11.1b

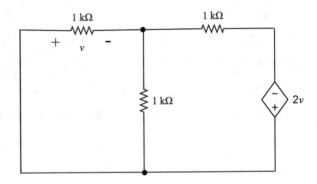

Case 2: Let us first remove the source on the right, obtaining Fig. P11.1c. The left two resistor are in parallel. Therefore, these two right resistors can be combined into a 0.5 k resistor, as shown in Fig. P11.1c.

Fig. P11.1c

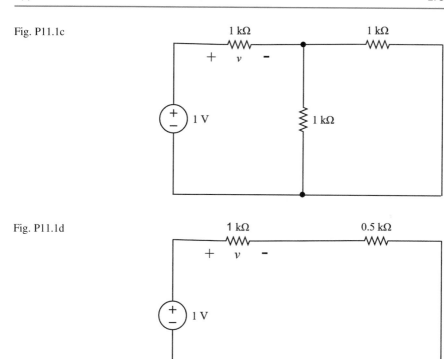

Fig. P11.1d

This is a voltage divider, and we have $v = 2/3$ V.

Combining the results from these two steps, the final answer is $v = 2/3$ V. However, the correct answer for v is 2 V.

Solution

The student's approach is wrong. We cannot treat a controlled source as an independent source. A controlled source can never be removed. This circuit has only one independent source, and we cannot use the super position principle to solve this problem.

Problem 11.2 Another student tries to use the superposition principle to solve for the voltage v in a different problem. The student's answer is **wrong**. Please help this student to find the mistake. Let us start with Fig. P11.2a.

Fig. P11.2a

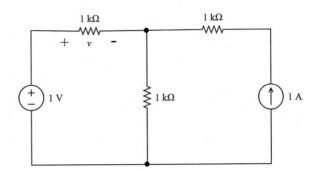

There are two independent sources in this circuit, and we will use the superposition principle to solve this problem.

Case 1: We remove the left source. The circuit becomes Fig. P11.2b.

Fig. P11.2b

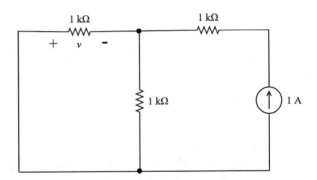

The 1 k resistor on the right has no effect in the circuit. The two resistors on the left is a current divider. Each 1 kΩ resistor on the left gets 0.5A of current. Using Ohm's law, $v = -500$ V.

Case 2: We remove the right source. The circuit becomes Fig. P11.2c.

Fig. P11.2c

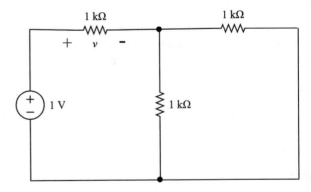

After combining the right two 1 kΩ resistors into a 0.5-kΩ resistor, we obtain a voltage divider. In this case, $v = 2/3$ V.

According to the superposition principle, we combine these two answers and obtain the final answer of

$$v = \frac{2}{3} - \frac{1}{2} = \frac{1}{6} \text{ V}.$$

However, the correct answer is

$$v = -499.5 \text{ V}.$$

What is wrong?

Solution

In the second case, when we remove the current source, we should leave the circuit open (instead of shorting the circuit).

Remember:
Removing a voltage source = setting the voltage to 0 = short.
Removing a current source = setting the current to 0 = open.

Problem 11.3 Use the superposition principle to solve for the voltage v. The circuit is

Fig. P11.3

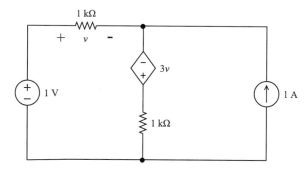

Solution
Case 1: Remove the left source by shorting the circuit and consider the circuit in Fig. S11.3a.

Fig. S11.3a

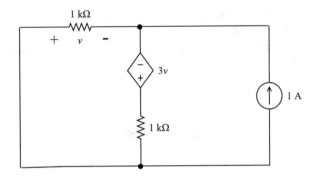

We can set up a node equation for this circuit.

$$\frac{-v}{1\text{ k}\Omega} + \frac{-v + 3v}{1\text{ k}\Omega} = 1\text{ A}.$$

Thus,

$$v = 1000\text{ V}.$$

Case 2: Remove the right source by leaving the circuit open and consider the circuit in Fig. S11.3b.

Fig. S11.3b

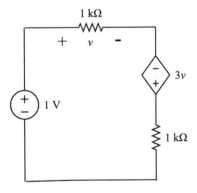

There is only one loop in the circuit, and the current is the same everywhere in the loop. The voltage rule applies. Each of the 1 kΩ resistor gets the same voltage. We have a KVL equation

$$v + v = 3v + (1\text{ V}).$$

Thus,

$$v = -1 \text{ V}.$$

Finally, combining results from Cases 1 and 2, we have

$$v = 1000 - 1 = 999 \text{ V}.$$

Chapter 12. Thévenin and Norton Equivalent Circuit

Problem 12.1 Find the Thévenin equivalent circuit of the circuit in Fig. P12.1.

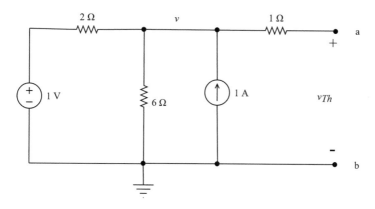

Fig. P12.1

Solution
To find R_{Th}, we remove all the independent sources in the circuit, obtaining Fig. S12.1.

The 2 Ω resistor and the 6 Ω resistor are in parallel, the combined resistor of them is 1.5 Ω. Therefore,

$$R_{\text{Th}} = (1 \ \Omega) + (1.5 \ \Omega) = 2.5 \ \Omega.$$

To find v_{Th}, we just need to use a node equation to find the voltage v as indicated in Fig. P12.1. Since there is no current in the 1 Ω resistor, $v_{\text{Th}} = v$. The node equation is as follows.

$$\frac{v - (1 \text{ V})}{2 \ \Omega} + \frac{v}{6 \ \Omega} = 1 \text{ A},$$

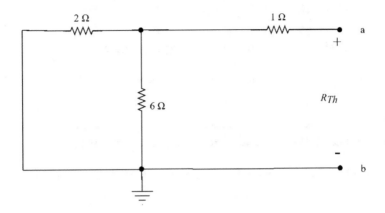

Fig. S12.1

where the 1 Ω resistor is not considered because there is no current through it. Solving this equation, we have

$$\frac{3v - (3 \text{ V})}{6\,\Omega} + \frac{v}{6\,\Omega} = 1 \text{ A},$$

$$4v = 9V,$$

$$v_{Th} = v = \frac{9}{4}\text{V}.$$

Problem 12.2 Find the Thévenin equivalent circuit of the circuit in Fig. P12.2.

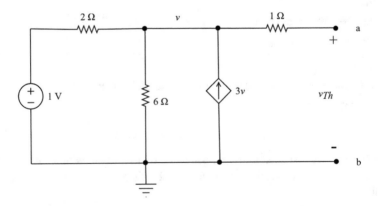

Fig. P12.2

Solution

To find v_{Th}, we just need to use a node equation to find the voltage v as indicated in Fig. P12.2. Since there is no current in the 1 Ω resistor, $v_{Th} = v$. The node equation is as follows.

$$\frac{v - (1 \text{ V})}{2 \, \Omega} + \frac{v}{6 \, \Omega} = 3v,$$

where the 1 Ω resistor is not considered because there is no current through it. Solving this equation, we have

$$\frac{3v - (3 \text{ V})}{6 \, \Omega} + \frac{v}{6 \, \Omega} = 3v,$$

$$3v - (3 \text{ V}) + v = 18v,$$

$$-14v = 3 \text{ V},$$

$$v_{Th} = v = -\frac{3}{14} \text{ V}.$$

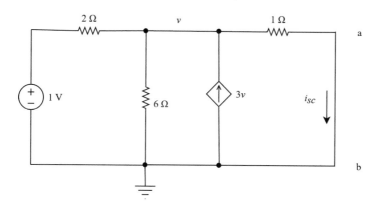

Fig. S12.2

This circuit contains a controlled source; we cannot find R_{Th}, by simply removing the independent sources. We use a different approach. We short circuit the output port and calculate the short-circuit current i_{sc} as labeled in Fig. S12.2.

We use the node equation to find the node voltage v and then use Ohm's law to find i_{sc}. The node equation for the circuit in Fig.S12.2 is given below.

$$\frac{v - (1\ \text{V})}{2\ \Omega} + \frac{v}{6\ \Omega} + \frac{v}{1\ \Omega} = 3v,$$

$$\frac{3v - (3\ \text{V})}{6\ \Omega} + \frac{v}{6\ \Omega} + \frac{6v}{6\ \Omega} = 3v,$$

$$3v - (3\ \text{V}) + v + 6v = 18v,$$

$$-8v = 3\ \text{V},$$

$$v = -\frac{3}{8}\ \text{V}.$$

Applying Ohm's law on the 1 Ω resistor,

$$i_{\text{sc}} = \frac{-\frac{3}{8}\ \text{V}}{1\ \Omega} = -\frac{3}{8}\ \text{A}.$$

Finally,

$$R_{\text{Th}} = \frac{v_{\text{Th}}}{i_{\text{sc}}} = \frac{-\frac{3}{14}\ \text{V}}{-\frac{3}{8}\ \text{A}} = \frac{4}{7}\ \Omega.$$

Problem 12.3 Use the testing source method to find the Thévenin resistance R_{Th} in Problem 12.2.

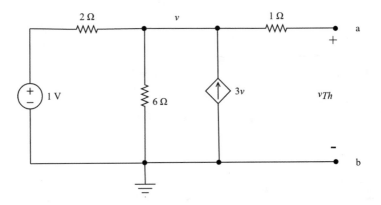

Fig. P12.3

Solution
We first need to remove the independent source and add a test source at the output port as shown in Fig. S12.3.

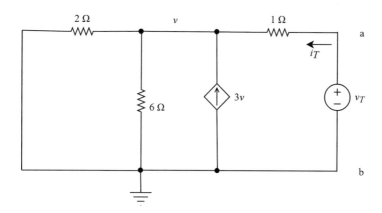

Fig. S12.3

Let us set up the node equation for the node voltage v.

$$\frac{v}{2\,\Omega} + \frac{v}{6\,\Omega} + \frac{v - v_T}{1\,\Omega} = 3v,$$

$$\frac{3v}{6\,\Omega} + \frac{v}{6\,\Omega} + \frac{6v - 6v_T}{6\,\Omega} = 3v,$$

$$3v + v + 6v - 6v_T = 18v,$$

$$-8v = 6v_T,$$

$$v = -\frac{3}{4}v_T.$$

Applying Ohm's law on the 1 Ω resistor,

$$i_T = \frac{v_T - v}{1\,\Omega} = v_T + \frac{3}{4}v_T = \frac{7}{4}v_T.$$

Finally,

$$R_{\text{Th}} = \frac{v_T}{i_T} = \frac{v_T}{\frac{7}{4}\, v_T} = \frac{4}{7}\ \Omega,$$

which is the same answer we obtain from Problem 12.2.

Problem 12.4 Find the Norton equivalent circuit of the circuit in Fig. P12.4.

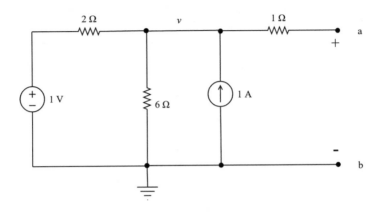

Fig. P12.4

Solution
This circuit is identical to that in Problem 12.1. The Norton resistance is the same as the Thévenin resistance. The methods for evaluating the Norton resistance are the same for evaluating the Thévenin resistance. Therefore, we can directly use the result from Problem 12.1 and the Norton resistance is

$$R_{\text{Nor}} = R_{\text{Th}} = 2.5\ \Omega.$$

If we already know the Thévenin voltage, the Norton current can be readily calculated as

$$i_{\text{Nor}} = \frac{v_{\text{Th}}}{R_{\text{Nor}}} = \frac{-\frac{9}{4}\ V}{\frac{5}{2}\ \Omega} = -\frac{9}{10}\ \text{A}.$$

If the Thévenin voltage is not available, the Norton current can be calculated by the short current at the output port (see Fig. S12.4).

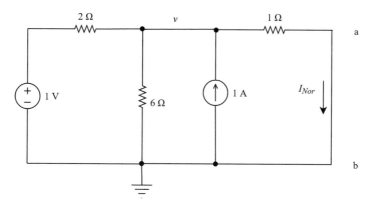

Fig. S12.4

The node equation for v is

$$\frac{v - (1\text{ V})}{2\,\Omega} + \frac{v}{6\,\Omega} + \frac{v}{1\,\Omega} = 1\text{ A},$$

$$\frac{3v - (3\text{ V})}{6\,\Omega} + \frac{v}{6\,\Omega} + \frac{6v}{6\,\Omega} = 1\text{ A},$$

$$10v = 9\text{ V},$$

$$v = \frac{9}{10}\text{ V}.$$

Using Ohm's law on the 1 Ω resistor yields

$$i_{\text{Nor}} = \frac{v}{1\,\Omega} = \frac{\frac{9}{10}\text{ V}}{1\,\Omega} = \frac{9}{10} = 0.9\text{ A}.$$

We reach the same answer.

Problem 12.5 Find the Norton equivalent circuit of the circuit in Fig. P12.5, using the step-by-step Thévenin/Norton conversion method.

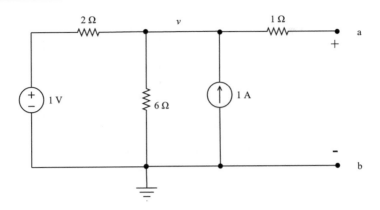

Fig. P12.5

Solution
In the first step, we convert the 1 V voltage source and the 2 Ω resistor into a Norton equivalence, as shown in Fig. S12.5a, where the current source value is obtained by (1 V)/(2 Ω) = 0.5 A.

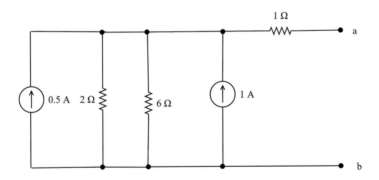

Fig. S12.5a

Combining the 2 Ω resistor and the 6 Ω resistor yields a 1.5-Ω resistor; combining the 1 A source and the 0.5 A source yields a 1.5-A source (See Fig. S12.5b).

Fig. S12.5b

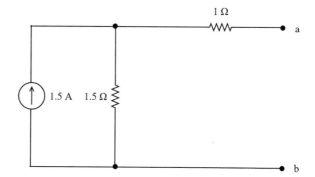

Next, we convert the 1.5 A current source and the 1.5 Ω resistor into a Thévenin equivalence, as shown in Fig. S12.5c, where the voltage source value is obtained by (1.5 A)(1.5 Ω) = 2.25 V.

Fig. S12.5c

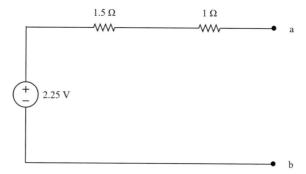

Combining the 1.5 Ω resistor and the 1 Ω resistor results in a 2.5-Ω resistor, as shown in Fig. S12.5d, which is the Thévenin equivalent circuit of Fig. P13.1.

Fig. S12.5d

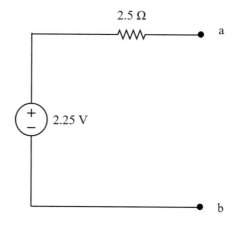

Finally, the Norton equivalent circuit is readily obtained from Fig. S12.5d as Fig. S12.5e, where the Norton current value is calculated by (2.25 V)/ (2.5 Ω) = 0.9 A.

Fig. S12.5e

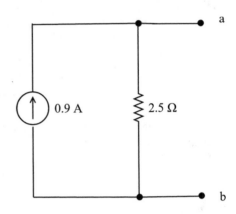

Chapter 13. Maximum Power Transfer

Problem 13.1 Find the maximum power delivered to R in the circuit in Fig. P13.1 when R is set for maximum power transfer?

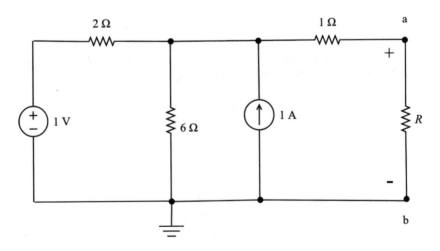

Fig. P13.1

Solution
From Problem 12.1, the Thévenin equivalent circuit has a Thévenin voltage of 2.25 V and a Thévenin resistance of 2.5 Ω (see Fig. S12.5d). Figure P13.1 is equivalent to Fig. S13.1.

Fig. S13.1

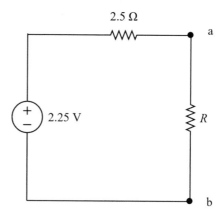

According to the maximum power transfer principle, when $R = R_{Th} = 2.5\ \Omega$, the load R receives the maximum power.

The maximum power received by the load in this case is

$$P_{Load} = \frac{(v_{Th}/2)^2}{R_{Th}} = \frac{((2.25\ \text{V})/2)^2}{2.5\ \Omega} = \frac{81}{160} = 0.50625\ \text{W}.$$

Problem 13.2 In Problem 13.1, let $R = 2.5\ \Omega$. What is the power provided by the 1 V voltage source? What is the power provided by the 1 A current source? In the circuit of Fig. P13.2, what percentage of the power delivered to the load $R = 2.5\ \Omega$ by the two sources?

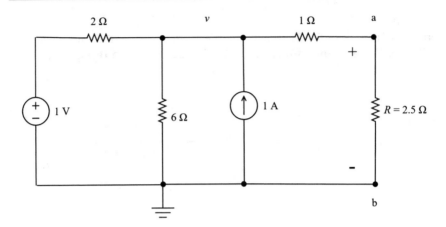

Fig. P13.2

Solution

To find the power of an individual power source, we have to use the original circuit, instead of the Thévenin equivalent circuit.

The node equation for v is

$$\frac{v - (1\text{ V})}{2\ \Omega} + \frac{v}{6\ \Omega} + \frac{v}{1\ \Omega + 2.5\ \Omega} = 1\text{ A},$$

$$\frac{21v - (21\text{ V})}{42\ \Omega} + \frac{7v}{42\ \Omega} + \frac{12v}{42\ \Omega} = 1\text{ A},$$

$$21v - (21\text{ V}) + 7v + 12v = 42\text{ V},$$

$$40v = 63\text{ V},$$

$$v = \frac{63}{40}\text{ V}.$$

Using Ohm's law, the current running through the 2 Ω resistor from right to left is

$$i_{2\Omega} = \frac{v - (1\text{ V})}{2\ \Omega} = \frac{\left(\frac{63}{40}\text{ V}\right) - (1\text{ V})}{2\ \Omega} = \frac{23}{80}\text{ A}.$$

This current is running into the positive terminal of the voltage source. This means that this voltage is not providing any power to the circuit. Instead, this voltage source is taking power from the circuit. In every life, this is the situation of a battery is being charged. The power that is delivered from this voltage source is

$$P_{\text{voltage source}} = -(1\text{ V})\left(\frac{23}{80}\text{ A}\right) = -\frac{23}{80}\text{ W}.$$

For the 1A current source, the current is flowing out and pointing to v, which is positive. This means that this source is providing power to the circuit. The voltage across the current source is $v = 63/40$ V. The power the is delivered from this voltage source is

$$P_{\text{current source}} = (1 \text{ A})\left(\frac{63}{40}\text{ V}\right) = \frac{63}{40}\text{ W}.$$

The current source gives 23/80 W to charge the voltage source and gives

$$\left(\frac{63}{40}\text{ W}\right) - \left(\frac{23}{80}\text{ W}\right) = \frac{103}{80}\text{ W}$$

to the rest of the circuit. This 103/80 W is the total delivered power by the sources. According to the result of Problem 13.1, the load consumes 81/160 W. The percentage of the power delivered to the load $R = 2.5\ \Omega$ by the two sources is

$$\frac{\frac{81}{160}}{\frac{103}{80}} = \frac{81}{206} = 39.3\%.$$

We notice that under the maximum power transfer condition that the load resistance equals to the Thevenin resistance, the percentage of the power delivered to the load may not be 50%.

Problem 13.3 As shown by the result of Problem 13.2, when the load resistance equals to the Thevenin resistance, the percentage of the power delivered to the load can be less than 50%. Use an example to explain this phenomenon.

Solution
Let us consider a circuit in Fig. S13.3a and its Thévenin equivalent circuit (Fig. S13.3b). The resistor R1 does not affect the Thévenin equivalent circuit.

Fig. S13.3a

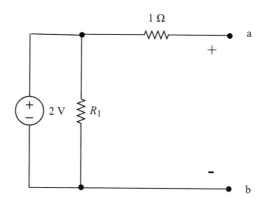

Fig. S13.3b

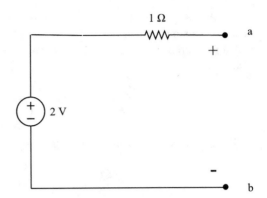

When a load of $R = 1\ \Omega$ is attached to the output port a-b, the load will get a maximum power of 1 W according to Fig. S13.3b.

We now calculate the power delivered from the 2 V voltage source. After a load of $R = 1\ \Omega$ is attached to the output port a-b. Using Ohm's law, the current running through the two $1\ \Omega$ resistors is

$$\frac{2\ \text{V}}{(1\ \Omega) + (1\ \Omega)} = 1\ \text{A}.$$

Also using Ohm's law, the current running through the resistor R_1 is

$$\frac{2\ \text{V}}{R_1}.$$

Therefore, the power delivered by the 2 V voltage source is

$$P_{\text{deliverd}} = \left[(1\ \text{A}) + \frac{2\ \text{V}}{R_1}\right] \times (2\ \text{V}),$$

which varies with the value of R_1. The percentage of the power delivered to the optimally matched load is given as

$$\frac{P_{\text{loadd}}}{P_{\text{deliverd}}} = \frac{1\ \text{W}}{\left[(1\ \text{A}) + \frac{2\ \text{V}}{R_1}\right] \times (2\ \text{V})}.$$

When $R_1 = \infty$,

$$\frac{P_{\text{loadd}}}{P_{\text{deliverd}}} = \frac{1\ \text{W}}{(1\ \text{A}) \times (2\ \text{V})} = \frac{1}{2} = 50\%.$$

When $R_1 = 2\ \Omega$,

$$\frac{P_{\text{loadd}}}{P_{\text{deliverd}}} = \frac{1 \text{ W}}{\left[(1 \text{ A}) + \frac{2 \text{ V}}{2 \Omega}\right] \times (2 \text{ V})} = \frac{1}{4} = 25\%.$$

Chapter 14. Operational Amplifiers

Problem 14.1 Express the output voltage v_{out} in terms of the inputs v_1 and v_2.

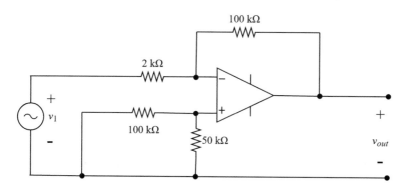

Fig. P14.1

Solution
This circuit has two inputs v_1 and v_2. We will use the superposition principle to find the output v_{out}.

Case 1: Let $v_2 = 0$. Figure P15.1 becomes Fig. S15.1.

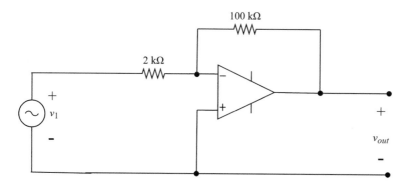

Fig. S14.1a

There is no current at the non-inverting input. Thus, there is no current in the resistors connecting to the non-inverting input. As a result, the voltage at the non-inverting input is zero. Figure S14.1a is then reduced to Fig. S14.1b.

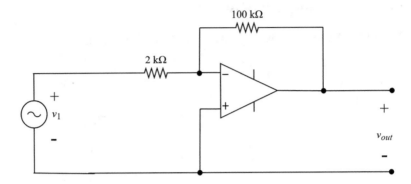

Fig. S14.1b

The circuit in Fig. S14.1b is an inverting amplifier. The output is evaluated as

$$v_{\text{out}} = -\frac{100 \text{ k}\Omega}{2 \text{ k}\Omega} v_1 = -50v_1.$$

Case 2: Let $v_1 = 0$. Figure P15.1 becomes Fig. S14.1c.

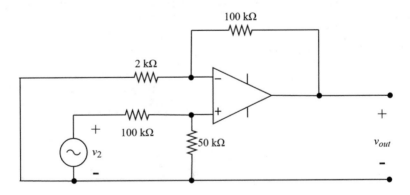

Fig. S14.1c

There is no current at the non-inverting input. The current in the two resistors connected to the non-inverting is the same; therefore, we can use the voltage divider method to find the voltage at the non-inverting input as

$$v_+ = \frac{50 \text{ k}\Omega}{(100 \text{ k}\Omega) + (50 \text{ k}\Omega)} v_2 = \frac{1}{3} v_2.$$

Figure S15.3a is a non-inverting amplifier. Once v_+ is known, the output can be readily evaluated as

$$v_{\text{out}} = \left(1 + \frac{100 \text{ k}\Omega}{2 \text{ k}\Omega}\right) v_+ = 51 v_+ = \frac{51}{3} v_2.$$

According to the principle of superposition, we combine the results from Cases 1 and 2, obtaining

$$v_{\text{out}} = \frac{51}{3} v_2 - 50 v_1.$$

Problem 14.2 Consider a current source as the inverting input. Find the current running into the output terminal.

Fig. P14.2

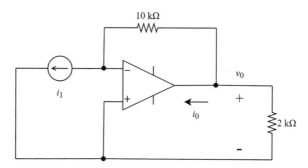

Solution
Since there is no current at the inverting input, the 10 kΩ resistor has the same current as i_1, running from right to left.

Using Ohm's law, we have

$$v_0 = (10 \text{ k}\Omega)(i_1).$$

The node equation for v_0 is

$$\frac{v_0}{2\,\text{k}\Omega} = i_1 + i_0.$$

Combining the above two equations yields

$$\frac{(10\,\text{k}\Omega)(i_1)}{2\,\text{k}\Omega} = i_1 + i_0,$$

$$5i_1 = i_1 + i_0,$$

$$i_0 = 4i_1.$$

Normally we are not interested in the current i_0 at all in practice.

If you treat the op-amp as a super node, we may find it strange that there is no current at the inverting input and the non-inverting input, but there is current at the output. To solve this paradox, we need to realize that the op-amp has three other terminals not commonly labeled in a circuit schematic: the positive power supply, the negative power supply, and the ground. The current running into the output terminal will go to the positive power supply, the negative power supply, or the ground.

Problem 14.3 The circuit shown in Fig. P14.3 can be thought of a current source. Find the range of the load R_L, in which the current in R_L is a constant. What is the value of this constant current?

Fig. P14.3

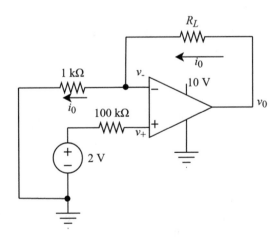

Solution

Since there is no current at the non-inverting input terminal,

$$v_+ = 2\,\text{V}.$$

Since there is no current at the inverting input terminal, the current in the 1 kΩ resistor and the load resistor R_L is the same. Thus, the voltage divider principle applies, and we have

$$v_- = \frac{1 \text{ k}\Omega}{R_L + (1 \text{ k}\Omega)} v_0.$$

Since

$$v_+ = v_-,$$

the current through the load R_L is expressed as

$$i_0 = \frac{v_-}{1 \text{ k}\Omega} = \frac{v_+}{1 \text{ k}\Omega} = \frac{2 \text{ V}}{1 \text{ k}\Omega} = 2 \text{ mA}.$$

This constant current $i_0 = 2$ mA is maintained when the output voltage v_0 in between the rail voltages

$$0 < v_0 < 10 \text{ V}.$$

Since

$$0 < v_0 = \frac{R_L + (1 \text{ k}\Omega)}{1 \text{ k}\Omega} v_- = \frac{R_L + (1 \text{ k}\Omega)}{1 \text{ k}\Omega} (2 \text{ V}) < 10 \text{ V},$$

$$R_L + (1 \text{ k}\Omega) < 5 \text{ k}\Omega.$$

The condition for constant current is

$$0 < R_L < 4 \text{ k}\Omega.$$

Chapter 15. Inductors

Problem 15.1 The switch closes at $t = 0$. Find the inductor current i_L as function of time.

Fig. P15.1

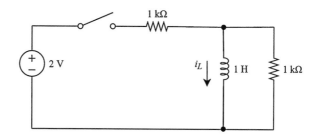

Solution

We only know how to solve a circuit that contain only one inductor and one resistor. Fig. P15.1 has two resistors!

To reduce this problem to a problem that we are able to solve, we treat the inductor as the circuit load and find the Thévenin equivalent circuit of Fig. P16.1, obtaining Fig. S15.1.

Fig. S15.1

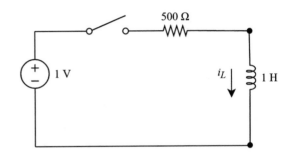

At the initial time $t = 0_-$,

$$i_L(0_-) = 0.$$

Since the inductor's current cannot suddenly change,

$$i_L(0_+) = 0.$$

At the final time $t = \infty$, the inductor can be treated as a conductor. Using Ohm's law,

$$i_L(\infty) = \frac{1 \text{ V}}{500 \text{ }\Omega} = 2 \text{ mA}.$$

The time constant is calculated as

$$\tau = \frac{L}{R} = \frac{1 \text{ H}}{500 \text{ }\Omega} = 0.002 \text{ s} = 2 \text{ ms}.$$

Using the general mathematical expression of the inductor current

$$i_L(t) = i_L(\infty) + [i_L(0) - i_L(\infty)]e^{-t/\tau}, \text{ for } t \geq 0,$$

we have

$$i_L(t) = 2 - 2e^{-t/(2\text{ms})} \text{ mA, for } t \geq 0.$$

Problem 15.2 We use the same circuit as in Problem 15.1. We assume that the switch has been closed for a long time. The switch opens at $t = 0$. Find the inductor current i_L as function of time.

Fig. P15.2

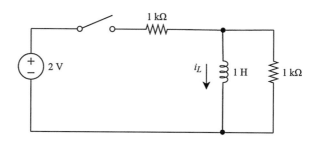

Solution

After the switch opens, only one resistor on the right is effective in the circuit, and the Thénenin equivalent circuit in Fig. S16.1 is no longer valid.

The initial condition of this problem is the final condition in Problem 15.1. Using the result of Problem 15.1, at the initial time $t = 0_-$,

$$i_L(0_-) = 2 \text{ mA}.$$

Since the inductor's current cannot suddenly change,

$$i_L(0_+) = 2 \text{ mA}.$$

At the final time $t = \infty$, the inductor can be treated a conductor and the current diminishes to 0. That is,

$$i_L(\infty) = 0.$$

The time constant is different from that in Problem 15.1 and is calculated as

$$\tau = \frac{L}{R} = \frac{1 \text{ H}}{1 \text{ k}\Omega} = 1 \text{ ms}.$$

Using the general mathematical expression of the inductor current

$$i_L(t) = i_L(\infty) + [i_L(0) - i_L(\infty)]e^{-t/\tau}, \text{for } t \geq 0,$$

we have

$$i_L(t) = 2e^{-t/(1\text{ms})} \text{ mA, for } t \geq 0.$$

Problem 15.3 The switch in the circuit in Fig. P16.2 has been closed for a long time before opening at $t = 0$. Find the inductor's current i_L and the inductor's voltage v_L for $t \geq 0$.

Fig. P15.3

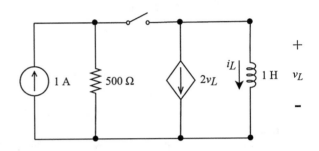

Solution
When the switch is closed for a long time, the inductor is a short circuit and the inductor voltage $v_L = 0$. As a result, the controlled (dependent) source is also 0. In other words, the dependent source is an open circuit. In this case, the inductor current is same as the independent source 1 A. Thus,

$$i_L(0_+) = i_L(0_-) = 1 \text{ A}.$$

At time $t = 0$, the switch opens, and Fig. P15.3 becomes Fig. S15.3a.

Fig. S15.3a

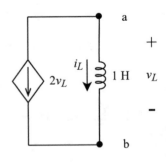

We do not know how to solve a circuit with an inductor and a dependent source. Fortunately, our old friend "Thévenin equivalent circuit" is able to help.

To find the Thévenin equivalent circuit for the dependent source in Fig. S16.3a, we apply a test source at the output port a-b, as indicated in Fig. S15.3b.

Fig. S15.3b

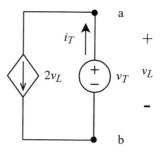

We have

$$i_T = 2v_L = 2v_T,$$

and the Thévenin resistance can be calculated as

$$R_{\text{Th}} = \frac{v_T}{i_T} = \frac{v_T}{2v_T} = 0.5 \ \Omega.$$

To find the Thévenin voltage, we need to first find the short-circuit current as shown in Fig. S15.3c.

Fig. S15.3c

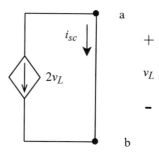

The short circuit leads to $v_L = 0$, which implies $2v_L = 0$ and $i_{\text{sc}} = 0$. Thus, the Thévenin voltage is

$$v_{\text{Th}} = i_{\text{sc}} \times R_{\text{Th}} = 0.$$

Using the Thévenin equivalent circuit, Fig. S16.3b becomes Fig. S15.3d.

Fig. S15.3d

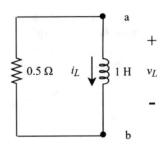

There is no source in Fig. S15.3d, and the inductor current will eventually diminish to 0. Thus,

$$i_L(\infty) = 0.$$

The time constant is calculated from Fig. S15.3d as

$$\tau = \frac{L}{R} = \frac{1\,\text{H}}{0.5\,\Omega} = 2\,\text{s}.$$

Using the general mathematical expression of the inductor current

$$i_L(t) = i_L(\infty) + [i_L(0) - i_L(\infty)]e^{-t/\tau}, \text{for } t \geq 0,$$

we have

$$i_L(t) = e^{-t/(1\text{s})} \text{ A, for } t \geq 0.$$

Finally, we find inductor voltage v_L for $t \geq 0$ according to the relationship

$$v_L(t) = L\frac{di_L(t)}{dt}.$$

Since $L = 1$ H, and we have

$$v_L(t) = \frac{de^{-t}}{dt} = -e^{-t} \text{ V, for } t \geq 0.$$

Chapter 16. Capacitors

Problem 16.1 The switch closes at $t = 0$. Find the inductor current i_L as function of time.

Fig. P16.1

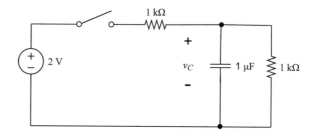

Solution
We only know how to solve a circuit that contain only one inductor and one resistor. Fig. P16.1 has two resistors!

To reduce this problem to a problem that we are able to solve, similar to Problem 15.1, we treat the capacitor as the circuit load and find the Thévenin equivalent circuit of Fig. P16.1, obtaining Fig. S16.1.

Fig. S16.1

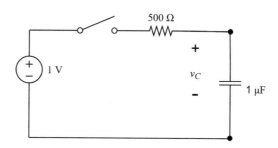

At the initial time $t = 0_-$,

$$v_C(0_-) = 0.$$

Since the capacitor's voltage cannot suddenly change,

$$v_C(0_+) = 0.$$

At the final time $t = \infty$, the capacitor can be treated an open circuit. Thus,

$$v_C(\infty) = 1 \text{ V}.$$

The time constant is calculated as

$$\tau = RC = (500 \ \Omega)(1 \ \mu F) = 0.5 \text{ ms}.$$

Using the general mathematical expression of the capacitor voltage

$$v_C(t) = v_C(\infty) + [v_C(0) - v_C(\infty)]e^{-t/\tau}, \text{for } t \geq 0,$$

we have

$$v_C(t) = 1 - e^{-t/(0.5\text{ms})} \text{ V, for } t \geq 0.$$

Problem 16.2 We use the same circuit as in Problem 16.1. We assume that the switch has been closed for a long time. The switch opens at $t = 0$. Find the capacitor's voltage v_C as function of time.

Fig. P16.2

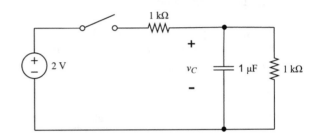

Solution

After the switch opens, only one resistor on the right is effective in the circuit, and the Thénenin equivalent circuit in Fig. S16.1 is no longer valid.

The initial condition of this problem is the final condition in Problem 16.1. Using the result of Problem 16.1, at the initial time $t = 0_-$,

$$v_C(0_-) = 1 \text{ V}.$$

Since the capacitor's voltage cannot suddenly change,

$$v_C(0_+) = 1 \text{ V}.$$

At the final time $t = \infty$, the capacitor is discharged to 0. Thus,

$$v_C(\infty) = 0.$$

The time constant is calculated as

$$\tau = RC = (1 \text{ k}\Omega)(1 \text{ μF}) = 1 \text{ ms}.$$

Using the general mathematical expression of the capacitor voltage

$$v_C(t) = v_C(\infty) + [v_C(0) - v_C(\infty)]e^{-t/\tau}, \text{for } t \geq 0,$$

we have

$$v_C(t) = 1 - e^{-t} \text{ V, for } t \geq 0.$$

Problem 16.3 The switch in the circuit in Fig. P16.3 has been closed for a long time before opening at $t = 0$. Find the capacitor's voltage v_C and the capacitor's current i_C for $t \geq 0$.

Solution

When the switch is closed for a long time, the capacitor is an open circuit and the

Fig. P16.3

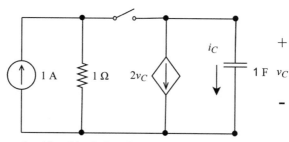

capacitor's voltage v_C can be solved by Ohm's law for the 1 Ω resistor.

$$v_C = (1 \text{ }\Omega)(1 - 2v_C),$$

$$3v_C = 1,$$

$$v_C = \frac{1}{3} \text{ V}.$$

This value is the voltage on the capacitor when the switch has been closed for a long time. At $t = 0$, the switch opens, we have

$$v_C(0_+) = v_C(0_-) = \frac{1}{3} \text{ V}.$$

After time $t = 0$, the switch opens, and Fig. P17.3 becomes Fig. S16.3a.

Fig. S16.3a

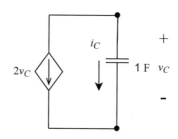

We do not know how to solve a circuit with an inductor and a dependent source. Fortunately, our old friend "Thévenin equivalent circuit" is able to help.

This dependent source is the same as that in Problem 15.3, Its Thévenin equivalent circuit for the dependent source is the same as that derived in Problem 15.3 and is nothing but a resistor of 0.5 Ω. Using the Thévenin equivalent circuit, we reach Fig. S16.3b.

Fig. S16.3b

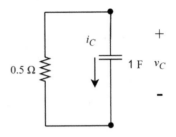

There is no source in Fig. S16.3b, and the capacitor's voltage will eventually diminish to 0. Thus,

$$v_C(\infty) = 0.$$

The time constant is calculated from Fig. S16.3b as

$$\tau = RC = (0.5\ \Omega)(1\ \text{F}) = 0.5\ \text{s}.$$

Using the general mathematical expression of the capacitor voltage

$$v_C(t) = v_C(\infty) + [v_C(0) - v_C(\infty)]e^{-t/\tau}, \text{for } t \geq 0,$$

we have

$$v_C(t) = \frac{1}{3}e^{-t/0.5}\ \text{V, for } t \geq 0.$$

Finally, we find capacitor's current i_C for $t \geq 0$ according to the relationship

$$i_C(t) = C\frac{dv_C(t)}{dt}.$$

Since $C = 1$ F, and we have

$$i_C(t) = \frac{1}{3}\frac{de^{-t/0.5}}{dt} = \frac{2}{3}e^{-t/0.5}\ \text{A, for } t \geq 0.$$

Chapter 17. Analysis of a Circuit by Solving Differential Equations

Problem 17.1 Set up a node equation for the circuit in Fig. P17.1. Then express the equation in terms of i_L.

Fig. P17.1

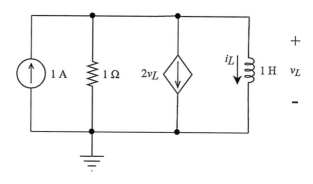

Solution
The node equation is

$$\frac{v_L}{1\,\Omega} = (1\ \text{A}) - (2v_L) - (i_L).$$

For the inductor, we have

$$v_L = L\frac{di_L}{dt} = (1\ \text{H})\frac{di_L}{dt},$$

and the node equation becomes a differential equation

$$\frac{di_L}{dt} = 1 - 2\frac{di_L}{dt} - i_L,$$

$$3\frac{di_L}{dt} + i_L - 1 = 0.$$

Problem 17.2 Set up a differential equation for $i_1 + i_2$.

Fig. P17.2

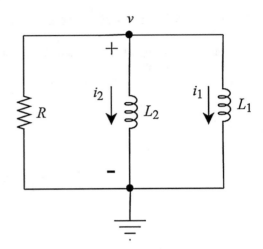

Solution

For indictors, the voltage and current follow the relationships

$$v(t) = L_1 \frac{di_1(t)}{dt} \text{ and } v(t) = L_2 \frac{di_2(t)}{dt},$$

$$\frac{1}{L_1} v(t) = \frac{di_1(t)}{dt} \text{ and } \frac{1}{L_2} v(t) = \frac{di_2(t)}{dt},$$

$$\left(\frac{1}{L_1} + \frac{1}{L_2} \right) v(t) = \frac{d[i_1(t) + i_2(t)]}{dt}.$$

The KCL equation for the circuit is

$$i_1(t) + i_2(t) + \frac{v(t)}{R} = 0,$$

$$i_1(t) + i_2(t) + \frac{L}{R} \frac{d[i_1(t) + i_2(t)]}{dt} = 0,$$

$$\frac{L}{R} \frac{di(t)}{dt} + i(t) = 0,$$

where

$$i(t) = i_1(t) + i_2(t),$$

$$\frac{1}{L} = \frac{1}{L_1} + \frac{1}{L_2}.$$

The differential equation

$$i_1(t) + i_2(t) + \frac{L}{R} \frac{d[i_1(t) + i_2(t)]}{dt} = 0$$

is solvable by solving

$$\frac{L}{R}\frac{di(t)}{dt} + i(t) = 0,$$

whose solution is

$$i(t) = i(\infty) + [i(0) - i(\infty)]e^{-Rt/L}, \text{ for } t \geq 0.$$

This circuit does not contain any sources, the combined inductor current $i(t)$ will eventually diminish to 0. That is,

$$i(t) = i(0)e^{-Rt/L}, \text{ for } t \geq 0.$$

This result $i(\infty) = 0$ does not imply $i_1(\infty) = i_2(\infty) = 0$. It only implies

$$i_1(\infty) = -i_2(\infty) = \text{constant}.$$

What is this constant? In fact, any constant will satisfy the differential equation. We need to find the constant satisfy the energy conservation.

The energy stored in an inductor is

$$w = \frac{1}{2}Li^2.$$

The initial energy of the system is determined by the initial currents in the inductors. The initial energy is, therefore,

$$w_0 = \frac{1}{2}L_1 i_1^2(0) + \frac{1}{2}L_2 i_2^2(0).$$

The power consumption of the resistor is

$$p_R = i^2 R = Ri^2(0)e^{-2Rt/L}, \text{ for } t \geq 0.$$

The total energy consumed by the resistor is

$$w_R = \int_0^\infty p_R dt = \int_0^\infty Ri^2(0)e^{-2Rt/L}dt = \frac{L}{2}i^2(0).$$

The final energy stored in the inductors is

$$w_\infty = \frac{1}{2}L_1 i_1^2(\infty) + \frac{1}{2}L_2 i_2^2(\infty) = \frac{i_1^2(\infty)}{2}(L_1 + L_2).$$

Energy conservation demands

$$w_\infty = w_0 - w_R,$$

$$\frac{i_1^2(\infty)}{2}(L_1 + L_2) = \frac{1}{2}L_1 i_1^2(0) + \frac{1}{2}L_2 i_2^2(0) - \frac{L}{2}i^2(0),$$

$$i_1^2(\infty) = \frac{L_1 i_1^2(0) + L_2 i_2^2(0) - L i^2(0)}{L_1 + L_2}.$$

In order to get some intuition, let us consider two numerical examples:

Example 1: $L_1 = L_2 = 1$ H and $i_1(0) = i_2(0) = 1$ A, we have $i_1(\infty) = -i_2(\infty) = 0$.
Example 2: $L_1 = L_2 = 1$ H, $i_1(0) = 1$ A and $i_2(0) = 0$, we have $i_1(\infty) = -i_2(\infty) = 0.5$ A.

This problem assumes ideal inductors, where the conductor is perfect with zero resistance. For an everyday inductor, we do not have $i(\infty) = 0.5$ A. The inductor current will eventually diminish and get $i(\infty) = 0$. The energy stored in the inductor is in the form of magnetic field.

Nowadays, the medical MRI machine uses liquid helium to create a superconducting environment so that the superconductor coil can maintain a strong magnetic field for a long time.

Problem 17.3 Set up a differential equation for $v_1 - v_2$.

Fig. P17.3

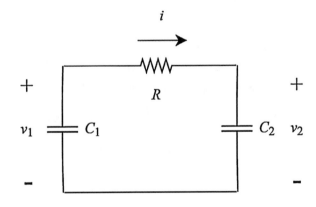

Solution
The KVL equation for the circuit is

$$v_1(t) - v_2(t) - R \times i(t) = 0.$$

For capacitors, the voltage and current follow the relationships

$$i(t) = C_2 \frac{dv_2(t)}{dt},$$

$$-i(t) = C_1 \frac{dv_1(t)}{dt}.$$

We have

$$-\frac{i(t)}{C_1} - \frac{i(t)}{C_2} = \frac{d[v_1(t) - v_2(t)]}{dt}.$$

Let

$$\frac{1}{C} = \frac{1}{C_1} + \frac{1}{C_1} \text{ and } v(t) = v_1(t) - v_2(t).$$

We have

$$-\frac{i(t)}{C_1} - \frac{i(t)}{C_1} = \frac{d[v_1(t) - v_2(t)]}{dt},$$

$$-i(t) = C\frac{d[v_1(t) - v_2(t)]}{dt},$$

$$[v_1(t) - v_2(t)] + RC\frac{d[v_1(t) - v_2(t)]}{dt} = 0,$$

$$v(t) + RC\frac{dv(t)}{dt} = 0.$$

The differential equation

$$v(t) + RC\frac{dv(t)}{dt} = 0$$

has a general solution with $\tau = RC$

$$v(t) = v(\infty) + [v(0) - v(\infty)]e^{-t/\tau}, \text{ for } t \geq 0.$$

Our circuit does not contain any sources, the combined capacitor voltage $v(t)$ will eventually diminish to 0. That is,

$$v(t) = v(0)e^{-t/\tau}, \text{ for } t \geq 0.$$

This result $v(\infty) = 0$ does not imply $v_1(\infty) = v_2(\infty) = 0$. It only implies

$$v_1(\infty) = v_2(\infty) = \text{constant}.$$

What is this constant? In fact, any constant will satisfy the differential equation. We need to find the constant satisfy the energy conservation.
The energy stored in a capacitor is

$$w = \frac{1}{2}Cv^2.$$

The initial energy of the system is determined by the initial voltages in the capacitors. The initial energy is, therefore,

$$w_0 = \frac{1}{2}C_1v_1^2(0) + \frac{1}{2}C_2v_2^2(0).$$

The power consumption of the resistor is

$$p_R = \frac{v^2}{R} = \frac{v^2(0)}{R}e^{-2t/\tau}, \text{ for } t \geq 0.$$

The total energy consumed by the resistor is

$$w_R = \int_0^\infty p_R dt = \int_0^\infty \frac{v^2(0)}{R}e^{-2t/\tau}dt = \frac{\tau}{2R}v^2(0) = \frac{C}{2}v^2(0).$$

The final energy stored in the inductors is

$$w_\infty = \frac{1}{2}C_1v_1^2(\infty) + \frac{1}{2}C_2v_2^2(\infty) = \frac{v_1^2(\infty)}{2}(C_1 + C_2).$$

Energy conservation demands

$$w_\infty = w_0 - w_R,$$

$$\frac{v_1^2(\infty)}{2}(C_1 + C_2) = \frac{1}{2}C_1v_1^2(0) + \frac{1}{2}C_2v_2^2(0) - \frac{C}{2}v^2(0),$$

$$v_1^2(\infty) = \frac{C_1v_1^2(0) + C_2v_2^2(0) - Cv^2(0)}{C_1 + C_2}.$$

In order to get some intuition, let us consider two numerical examples:

Example 1: $C_1 = C_2 = 1$ F and $v_1(0) = v_2(0) = 1$ V, we have $v_1(\infty) = v_2(\infty) = 0$.
Example 2: $C_1 = C_2 = 1$ F, $v_1(0) = 1$ V and $v_2(0) = 0$, we have $v_1(\infty) = v_2(\infty) = 0.5$ V.

This problem assumes ideal capacitors, where the dielectric between the two metal plates is perfect insulator. For an everyday capacitor, we do not have $v(\infty) = 0.5$ V. The capacitor will be eventually discharged and get $v(\infty) = 0$. The energy stored in the capacitor can be expressed in voltage (v) or in charge (Q) as

$$w = \frac{Cv^2}{2} = \frac{Q^2}{2C}.$$

Nowadays, the supercapacitors can store electric charges and used as batteries, but their self-discharge rate is significantly faster than rechargeable batteries.

Chapter 18. First-Order Circuits

Problem 18.1 The input of an RC circuit is a periodic square pulse sequence. The period is $2\ T$. The time constant of the RC circuit is τ. The output signal is the capacitor voltage v_C. Find the output signal's maximum value v_{max} and the minimum value v_{min}.

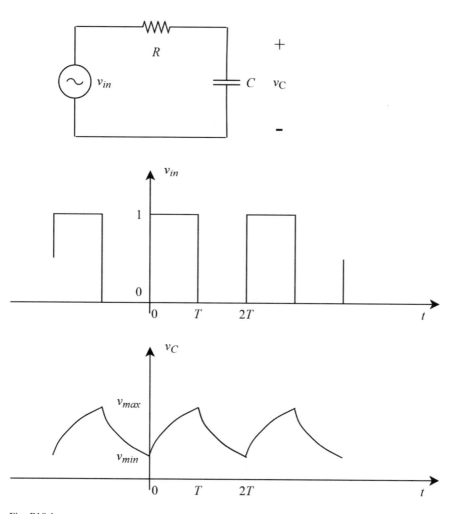

Fig. P18.1

Solution
This is a sequential switching first-order system. We will use this general solution twice.

$$v_C(t) = v_C(\infty) + [v_C(0) - v_C(\infty)]e^{-t/\tau}, \text{for } t \geq 0,$$

where $t = 0$ is understood as the initial time.

For the charging period,

$$v_C(0) = v_{\min},$$

$$v_C(\infty) = 1,$$

$$v_C(T) = v_{\max},$$

$$v_C(T) = 1 + [v_{\min} - 1]e^{-T/\tau} = v_{\max}.$$

For the discharging period,

$$v_C(0) = v_{\max},$$

$$v_C(\infty) = 0,$$

$$v_C(T) = v_{\min},$$

$$v_C(T) = v_{\max}e^{-T/\tau} = v_{\min}.$$

From

$$\begin{cases} 1 + [v_{\min} - 1]e^{-T/\tau} = v_{\max} \\ v_{\max}e^{-T/\tau} = v_{\min} \end{cases}$$

we obtain

$$\begin{cases} v_{\max} = \dfrac{1 - e^{-T/\tau}}{1 - e^{-2T/\tau}}, \\ v_{\min} = \dfrac{e^{T/\tau} - 1}{e^{2T/\tau} - 1}. \end{cases}$$

If the time constant τ is much longer than T, we have

$$\begin{cases} v_{\max} \approx 0.5, \\ v_{\min} \approx 0.5. \end{cases}$$

This circuit can be used to smooth the input signal. As we will find out later in this book, this circuit is a lowpass filter.

Problem 18.2 A student tries to solve a problem in his own way, and he does not get the correct answer. Please help him to find the error. In the problem, the switch has been closed for a long time. The switch opens at $t = 0$. Find the capacitor's voltage i_C as function of time.

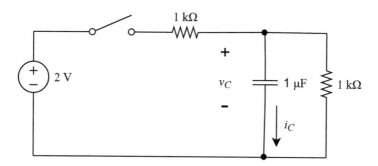

Fig. P18.2

The student's solution:
After the switch has been closed for a long time, the capacitor acts like an open circuit. Therefore,

$$i_C(0) = 0.$$

At $t = 0$, the switch opens. Now the circuit does not have any source, and the capacitor will eventually discharge to 0. Thus,

$$i_C(\infty) = 0.$$

Recall the general solution

$$i(t) = i(\infty) + [i(0) - i(\infty)]e^{-t/\tau}, \text{ for } t \geq 0.$$

The student's solution is

$$i_C(t) = 0, \text{ for } t \geq 0.$$

Solution
For the RC circuit, we need to solve the capacitor's voltage first. The capacitor's voltage has a general solution

$$v_C(t) = v_C(\infty) + [v_C(0) - v_C(\infty)]e^{-t/\tau}, \text{ for } t \geq 0.$$

After that capacitor's voltage $v_C(t)$ is found, we can use $v_C(t)$ to find other unknowns. For example, if we want to find the capacitor's current $i_C(t)$, we need to use the formula

$$i_C(t) = C \frac{dv_C(t)}{dt}.$$

If we want to find the voltage across the resistor, we use $v_C(t)$ and KVL. The voltage across a capacitor cannot change suddenly; however, the current through a capacitor can change suddenly. In an RC circuit, the voltage across the capacitor determines the behavior of the entire circuit.

Likewise, for the RL circuit, we need to solve the inductor's current first. The inductor's current has a general solution

$$i_L(t) = i_L(\infty) + [i_L(0) - i_L(\infty)]e^{-t/\tau}, \text{ for } t \geq 0,$$

After that inductor's current $i_L(t)$ is found, we can use $i_L(t)$ to find other unknowns. For example, if we want to find the inductor's voltage $v_L(t)$, we need to use the formula

$$v_L(t) = L \frac{di_L(t)}{dt}.$$

If we want to find the current through the resistor, we use $i_L(t)$ and KCL. The current through an inductor cannot change suddenly; however, the voltage across an inductor can change suddenly. In an RL circuit, the current through the inductor determines the behavior of the entire circuit.

Chapter 19. Sinusoidal Steady State

Problem 19.1 Express the following signals in the phasor form:

(a) $5 \cos (100t)$
(b) $5 \sin (100t)$
(c) $5 \cos \left(100t + 45^\circ\right)$
(d) $5 \sin \left(100t + 45^\circ\right)$
(e) $2 \cos (\omega t)$
(f) $2 \cos (\omega t) - 3 \cos (2\omega t)$
(g) 10
(h) $2t^2 \cos (\omega t)$

Solution

(a) $5 \angle 0°$

(b) $5 \sin (100t) = 5 \cos (100t - 90°)$, so, the phasor form is $5 \angle -90°$.

(c) $5\angle45°$

(d) $5 \sin (100t + 45°) = 5 \cos (100t + 45° - 90°)$, so, the phasor form is $5 \angle -45°$.

(e) 5

(f) $2 \cos (\omega t) - 3 \cos (2\omega t)$. This expression has two different frequencies. Cannot use phasor forms.

(g) This is a dc signal. Do not use sinusoidal analysis.

(h) Not a sinusoidal signal due to the $2t^2$ time-varying amplitude.

Problem 19.2 Express the transfer function in the phasor form. The input is v_{in} and the output is v_C.

Fig. P19.2

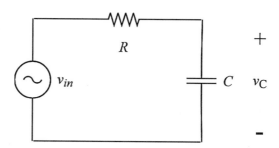

Solution

This is a voltage divider, and the capacitor can be treated as a "resistor" with impedance (similar to "resistance")

$$Z = \frac{1}{j\omega C} = \frac{1}{\omega C} \angle -90°.$$

The transfer function is the ratio of output over input, which is, in the phasor form,

$$H(\omega) = \frac{Z}{R + Z},$$

$$= \frac{\frac{1}{j\omega C}}{R + \frac{1}{j\omega C}},$$

$$= \frac{1}{j\omega CR + 1},$$

$$= \frac{1}{\sqrt{(\omega CR)^2 + 1} \angle \tan^{-1}(\omega CR)},$$

$$= \frac{1}{\sqrt{(\omega CR)^2 + 1}} \angle - \tan^{-1}(\omega CR).$$

You may wonder why we are interested in a transfer function. A transfer function comes in handy if you know the input and want to find the output.

Here is a numerical solution.

Let $R = 1$ kΩ, $C = 1$ μF, and $v_{in}(t) = 10 \cos (1000t)$. Thus, $\omega = 1000$ and the input phasor is just

$$10\angle 0^\circ.$$

In fact, when we convert the time-domain signal $v_{in}(t) = 10 \cos (1000t)$ to the phasor

$$10\angle 0^\circ = 10\cos\left(0^\circ\right) + j10\sin\left(0^\circ\right),$$

we have done two things. First, we discard the frequency $\omega = 1000$. Second, we add an imaginary term.

The output phasor in the product of the input phasor and the transfer function

$$\left(10\angle 0^\circ\right) \times \left[\frac{1}{\sqrt{(\omega CR)^2 + 1}} \angle - \tan^{-1}(\omega CR)\right]$$

$$= \left(10\angle 0^\circ\right) \times \left(\frac{1}{\sqrt{2}} \angle - \tan^{-1}(1)\right),$$

$$= \left(10\angle 0^\circ\right) \times \left(\frac{1}{\sqrt{2}} \angle - 45^\circ\right),$$

$$= \frac{10}{\sqrt{2}} \angle - 45^\circ.$$

You need to remind yourself that a phasor is just a complex number, which can be in the Cartesian form

$$a + jb$$

or in the polar form (exponential form)

$$\sqrt{a^2 + b^2} \angle \tan^{-1}\left(\frac{b}{a}\right) = \sqrt{a^2 + b^2}\, e^{\, j \tan^{-1}\left(\frac{b}{a}\right)}.$$

The product of two phasors is the product of two complex numbers. In the polar form, the magnitude is the product of the two magnitudes and the phase is the sum of the phases.

It is easier to use the Cartesian form to add and subtract phasors. It is easier to use the polar form (or exponential form) to multiply and divide phasors.

In the regular time-domain expression, the output signal is

$$v_C(t) = \frac{10}{\sqrt{2}} \cos\left(1000t - 45^\circ\right).$$

We must remember two things when converting a phasor back to the time-domain signal. First, we take real part (i.e., keeping the cosine function) and discard the imaginary part (i.e., removing the sine function). Second, insert the frequency $\omega = 1000$ in the cosine function as ωt. This frequency ω must be the same frequency ω in the input signal.

Chapter 20. Function Generators and Oscilloscopes

Problem 20.1 In Fig. P20.1, an oscilloscope is directly connected to a signal generator to verify the signal generated. We set the peak-to-peak voltage of a sinewave to be 10 V. However, the oscilloscope shows a 20-V peak-to-peak. Is there anything wrong?

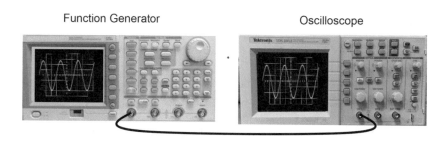

Function Generator Oscilloscope

Fig. P20.1

Solution
Nothing is wrong. This issue has already been discussed in text. We will discuss it again here.

The default setting for most function generators is to display the desired voltage as though terminated into a 50-Ohm load. When a high impedance device, such as an oscilloscope is used to measure the output of the function generator, the waveform appears to be twice the voltage set on the display of the oscilloscope.

Fig. S20.1

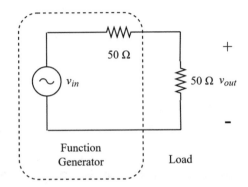

The Thévenin equivalent circuit for a typical signal generator is shown in Fig. S20.1. The value of v_{out} is displayed on the generator. If load is 50 Ω, the actual v_{in} is twice as large as v_{out}. Even if the load is large, the generator's display ignores the actual v_{out} and displays a value that is half of the value of v_{in}.

The input impedance of an oscilloscope is in the order of 1 MΩ; therefore, the oscilloscope shows almost a correct amplitude of v_{in}.

Problem 20.2 How to use an oscilloscope to estimate the time constant of a first-order circuit?

Solution
If your first-order circuit is an *RC* circuit, you can use an oscilloscope to measure the voltage across the capacitor.

If your first-order circuit is an *RL* circuit, do not use an oscilloscope to measure the voltage across the inductor because this voltage is almost zero. In this case, find a resistor in the circuit and measure the voltage drop across this resistor.

A portion of your signal is described by an exponential function

$$[v(0) - v(\infty)]e^{-t/\tau}.$$

This may be a charging or discharging trend, as shown in Fig. S20.2.

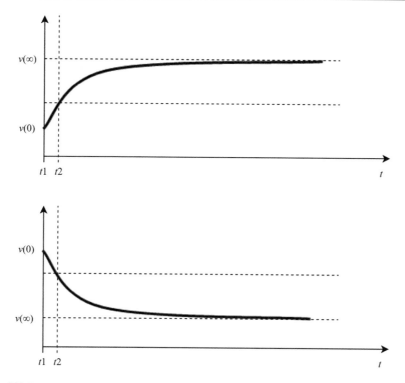

Fig. S20.2

Let $t = \tau$, and $e^{-t/\tau} = e^{-1} = 0.368$. Find the value that is

$$v_1 = 0.368 \times |v(\infty) - v(0)|.$$

Adjust the oscilloscope's vertical and time scaling knobs so that you see a charging or discharging period nice and big.

Let us assume that you are looking at the charging period.

When you hit the cursor button on the oscilloscope, a menu should come up on the screen saying "Cursors." Set the time position of the cursor at the point the charging begins. This is your $t1$. Then move the time position of the cursor to position the curve, that is v_1, above the starting position. This is your $t2$. Your time constant is

$$\tau = t2 - t1.$$

Chapter 21. Mutual Inductance and Transformers

Problem 21.1 An ideal transformer has 1000 turns in its primary coil and 100 turns in its secondary coil. Determine whether the following statements are true.

(a) This is a 10:1 transformer.
(b) This is a 1:10 transformer.
(c) This is a 1:0.1 transformer.
(d) This is a 0.1:1 transformer.
(e) This is a step-up transformer.
(f) This is a step-down transformer.
(g) If the primary voltage is 10 V, the secondary voltage is 100 V.
(h) If the primary voltage is 10 V, the secondary voltage is 1 V.
(i) If the primary current is 10 A, the secondary current is 100 A.
(j) If the primary current is 10 A, the secondary current is 1 A.
(k) The transformer consumes power.
(l) The transformer only works for a DC source input.
(m) The transformer only works for an AC source input.
(n) The frequency on secondary side is 10 times higher than the frequency on the primary side.
(o) The frequency on secondary side is 10 times lower than the frequency on the primary side.
(p) If the secondary side has a load of 100 Ω, the reflected impedance on the primary side is 1000 Ω.
(q) If the secondary side has a load of 100 Ω, the reflected impedance on the primary side is 10,000 Ω.

Solution
(a) This is a 10:1 transformer. This is true.
(b) This is a 1:10 transformer. This is false.
(c) This is a 1:0.1 transformer. This is true.
(d) This is a 0.1:1 transformer. This is false.
(e) This is a step-up transformer. This is false.
(f) This is a step-down transformer. This is true.
(g) If the primary voltage is 10 V, the secondary voltage is 100 V. This is false.
(h) If the primary voltage is 10 V, the secondary voltage is 1 V. This is true.
(i) If the primary current is 10 A, the secondary current is 100 A. This is true.
(j) If the primary current is 10 A, the secondary current is 1 A. This is false.
(k) The transformer consumes power. This is false.
(l) The transformer only works for a DC source input. This is false.
(m) The transformer only works for an AC source input. This is true.
(n) The frequency on secondary side is 10 times higher than the frequency on the primary side. This is false. The frequency does not change.
(o) The frequency on secondary side is 10 times lower than the frequency on the primary side. This is false.

(p) If the secondary side has a load of 100 Ω, the reflected impedance on the primary side is 1000 Ω. This is false.

(q) If the secondary side has a load of 100 Ω, the reflected impedance on the primary side is 10,000 Ω. This is true.

Problem 21.2 This problem is about the dot notation and convention in a transformer. Express the induced voltages for each case.

(a)

Fig. P21.2a
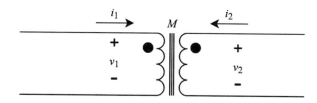

(b)

Fig. P21.2b
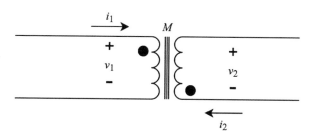

(c)

Fig. P21.2c
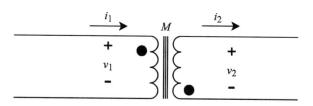

(d)

Fig. P21.2d
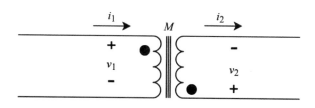

(e)

Fig. P21.2e

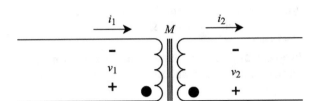

(f)

Fig. P21.2f

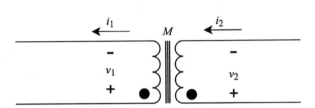

(g)

Fig. P21.2g

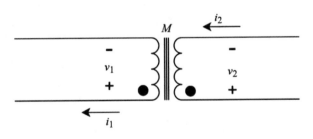

(h)

Fig. P21.2h

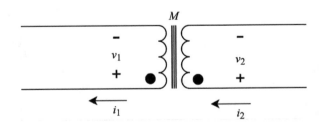

Solution

(a) $v_2 = M\frac{di_1}{dt}$ and $v_1 = M\frac{di_2}{dt}$.

(b) $v_2 = -M\frac{di_1}{dt}$ and $v_1 = M\frac{di_2}{dt}$.

(c) $v_2 = -M \frac{di_1}{dt}$ and $v_1 = M \frac{di_2}{dt}$.

(d) $v_2 = -M \frac{di_1}{dt}$ and $v_1 = M \frac{di_2}{dt}$.

(e) $v_2 = M \frac{di_1}{dt}$ and $v_1 = M \frac{di_2}{dt}$.

(f) $v_2 = M \frac{di_1}{dt}$ and $v_1 = -M \frac{di_2}{dt}$.

(g) $v_2 = -M \frac{di_1}{dt}$ and $v_1 = -M \frac{di_2}{dt}$.

(h) $v_2 = -M \frac{di_1}{dt}$ and $v_1 = M \frac{di_2}{dt}$.

Chapter 22. Fourier Series

Problem 22.1 Match a Fourier series with a periodic function with $\omega_0 = 2\pi/T$.

$$f_1(t) = a_0 + \sum_{n=1}^{\infty} a_n \cos(n\omega_0 t)$$

$$f_2(t) = \sum_{n=1}^{\infty} a_n \cos(n\omega_0 t)$$

$$f_3(t) = a_0 + \sum_{n=1}^{\infty} b_n \sin(n\omega_0 t)$$

$$f_4(t) = \sum_{n=1}^{\infty} b_n \sin(n\omega_0 t)$$

$$f_5(t) = \sum_{\substack{n=1 \\ n=\text{odd}}}^{\infty} [a_n \cos(n\omega_0 t) + b_n \sin(n\omega_0 t)]$$

$$f_6(t) = \sum_{\substack{n=1 \\ n=\text{odd}}}^{\infty} a_n \cos(n\omega_0 t)$$

$$f_7(t) = \sum_{\substack{n=1 \\ n=\text{odd}}}^{\infty} b_n \sin(n\omega_0 t)$$

$$f_8(t) = a_0 + \sum_{n=1}^{\infty} [a_n \cos(n\omega_0 t) + b_n \sin(n\omega_0 t)]$$

$$f_9(t) = \sum_{n=1}^{\infty} [a_n \cos(n\omega_0 t) + b_n \sin(n\omega_0 t)]$$

(a)

Fig. P22.1a

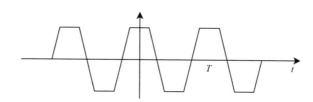

(b)

Fig. P22.1b

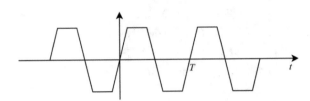

(c)

Fig. P22.1c

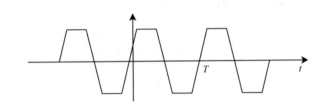

(d)

Fig. P22.1d

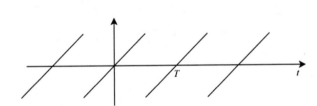

(e)

Fig. P22.1e

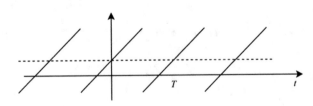

(f)

Fig. P22.1f

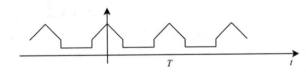

(g)

Fig. P22.1g

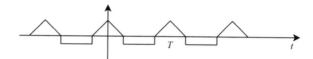

(h)

Fig. P22.1h

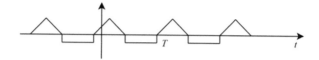

(i)

Fig. P22.1i

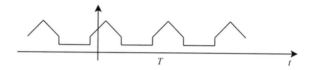

Solution

(a) $f_6(t) = \sum_{n=1}^{\infty} a_n \cos(n\omega_0 t)$

$\qquad\qquad n = \text{odd}$

(b) $f_7(t) = \sum_{n=1}^{\infty} b_n \sin(n\omega_0 t)$

$\qquad\qquad n = \text{odd}$

(c) $f_5(t) = \sum_{n=1}^{\infty} [a_n \cos(n\omega_0 t) + b_n \sin(n\omega_0 t)]$

$\qquad\qquad n = \text{odd}$

(d) $f_4(t) = \sum_{n=1}^{\infty} b_n \sin(n\omega_0 t)$

(e) $f_3(t) = a_0 + \sum_{n=1}^{\infty} b_n \sin(n\omega_0 t)$

(f) $f_1(t) = a_0 + \sum_{n=1}^{\infty} a_n \cos(n\omega_0 t)$

(g) $f_2(t) = \sum_{n=1}^{\infty} a_n \cos(n\omega_0 t)$

(h) $f_9(t) = \sum_{n=1}^{\infty} [a_n \cos(n\omega_0 t) + b_n \sin(n\omega_0 t)]$

(i) $f_8(t) = a_0 + \sum_{n=1}^{\infty} [a_n \cos(n\omega_0 t) + b_n \sin(n\omega_0 t)]$.

Problem 22.2 Show that the Fourier series has an equivalent exponential form

$$f(t) = \sum_{n=-\infty}^{\infty} c_n e^{jn\omega_0 t}.$$

Solution

Using Euler's formula

$$e^{jn\omega_0 t} = \cos(n\omega_0 t) + j\sin(n\omega_0 t),$$

We have

$$\cos(n\omega_0 t) = \frac{e^{jn\omega_0 t} + e^{-jn\omega_0 t}}{2},$$

$$\sin(n\omega_0 t) = \frac{e^{jn\omega_0 t} - e^{-jn\omega_0 t}}{2j}.$$

The Fourier series

$$f(t) = a_0 + \sum_{n=1}^{\infty} [a_n \cos(n\omega_0 t) + b_n \sin(n\omega_0 t)]$$

can be rewritten as

$$f(t) = a_0 + \sum_{n=1}^{\infty} \left[a_n \frac{e^{jn\omega_0 t} + e^{-jn\omega_0 t}}{2} + b_n \frac{e^{jn\omega_0 t} - e^{-jn\omega_0 t}}{2j} \right],$$

$$f(t) = a_0 + \sum_{n=1}^{\infty} \left[\frac{a_n - jb_n}{2} e^{jn\omega_0 t} + \frac{a_n + jb_n}{2} e^{-jn\omega_0 t} \right],$$

$$f(t) = c_0 + \sum_{n=1}^{\infty} \left[c_n e^{jn\omega_0 t} + c_{-n} e^{-jn\omega_0 t} \right].$$

where

$$c_n = \frac{a_n - b_n}{2},$$

$$c_0 = a_0.$$

Chapter 23. Laplace Transform in Circuit Analysis

Problem 23.1 Solve the following differential equation using the Laplace transform method.

$$x''(t) + 4x'(t) + 3x(t) = 5$$

with initial conditions $x'(0) = 1$ and $x(0) = 2$.

Solution
Taking the Laplace transform term by term according to the Laplace transform table, we have

$$s^2X(s) - sx(0) - x'(0) + 4sX(s) - 4x(0) + 3X(s) = \frac{5}{s},$$

$$s^2X(s) - 2s - 1 + 4sX(s) - 8 + 3X(s) = \frac{5}{s},$$

$$X(s) = \frac{\frac{5}{s} + 9 + 2s}{s^2 + 4s + 3} = \frac{5 + 9s + 2s^2}{s(s+1)(s+3)} = \frac{k_1}{s} + \frac{k_2}{s+1} + \frac{k_3}{s+3}.$$

We can use the cover-up method to find the partial fraction expansion.

$$k_1 = \frac{5 + 9(0) + 2(0)^2}{(0+1)(0+3)} = \frac{5}{3},$$

$$k_2 = \frac{5 + 9(-1) + 2(-1)^2}{(-1)(-1+3)} = 1,$$

$$k_3 = \frac{5 + 9(-3) + 2(-3)^2}{(-3)(-3+1)} = -\frac{2}{3}.$$

The Laplace transform of the solution is

$$X(s) = \frac{5/3}{s} + \frac{1}{s+1} + \frac{-2/3}{s+3}.$$

The solution can be obtained by finding the inverse Laplace transform (using the Table of Laplace Transform Pairs):

$$x(t) = \frac{5}{3} + e^{-t} - \frac{2}{3}e^{-3t}, \text{ for } t \geq 0.$$

Problem 23.2 Use the Laplace transform method to solve for the circuit in Fig. P23.2. The initial voltage in the capacitor is 3 V.

Fig. P23.2

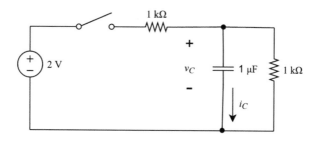

Solution

We first need to convert the time-domain circuit in Fig. P23.2 to its corresponding Laplace-domain circuit in Fig. S23.2a. The initial condition is converted into a voltage source or a current source.

We know that if the initial condition is zero, the time-domain relationship for the capacitor

$$i(t) = C \frac{dv(t)}{dt}$$

has a Laplace-domain counterpart

$$I(s) = CsV(s).$$

According to the Laplace transform table, if the initial condition is not zero, the Laplace-domain counterpart is

$$I(s) = C[sV(s) - v(0)] = CsV(s) - Cv(0) = Cs\left[V(s) - \frac{v(0)}{s}\right].$$

There are two ways to handle this initial condition. The first way is to use the relationship

$$I(s) = CsV(s) - Cv(0)$$

and treat the initial condition as a current source of value

$$-Cv(0) = -3\,\mu.$$

The second way is to use the relationship

$$V(s) = \frac{I(s)}{Cs} + \frac{v(0)}{s}$$

and treat the initial condition as a voltage source of value

$$\frac{v(0)}{s} = \frac{3}{s}.$$

Therefore, we can have two Laplace-main circuits, as shown in Fig. S23.2a and Fig. S23.2b, respectively.

Fig. S23.2a

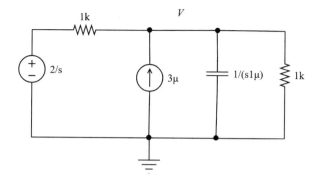

Fig. S23.2b

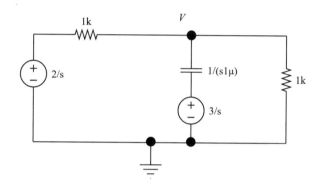

Let us first solve the circuit in Fig. S24.1. The node equation is

$$\frac{V - \frac{2}{s}}{1\,k} + \frac{V}{\frac{1}{s\,\mu}} + \frac{V}{1\,k} = 3\,\mu,$$

$$V = \frac{3\,\mu + \frac{2}{s\,1\,k}}{\frac{2}{1\,k} + s\,\mu} = \frac{3s + 2000}{s(2000 + s)} = \frac{k_1}{s} + \frac{k_2}{s + 2000}.$$

Using the cover-up method to find k_1 and k_2,

$$k_1 = \frac{3(0) + 2000}{2000 + 0} = 1,$$

$$k_2 = \frac{3(-2000) + 2000}{(-2000)} = 2.$$

Therefore,

$$V = \frac{1}{s} + \frac{2}{s + 2000}$$

and

$$v(t) = 1 + 2e^{-2000t} \; V, \text{for } t \geq 0.$$

Now let us solve the circuit in Fig. S23.2b. The node equation is

$$\frac{V - \frac{2}{s}}{1\,k} + \frac{V - \frac{3}{s}}{\frac{1}{s\,\mu}} + \frac{V}{1\,k} = 0,$$

$$V = \frac{\frac{2000}{s} + 3}{2000 + s} = \frac{3s + 2000}{s(2000 + s)}.$$

This expression is exactly the same as that for Fig. S23.2a. Therefore, we have the same solution

$$v(t) = 1 + 2e^{-2000t} \; V, \text{for } t \geq 0.$$

Problem 23.3 Use the Laplace transform method to solve for the circuit in Fig. P23.3. The initial current in the inductor is 3 A.

Fig. P23.3

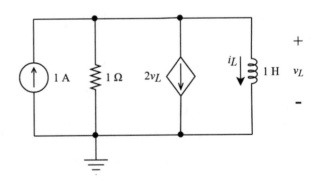

Solution
We first need to convert the time-domain circuit in Fig. P23.3 to its corresponding Laplace-domain circuit in Fig. S23.3a. The initial condition is converted into a voltage source or a current source.

We know that if the initial condition is zero, the time-domain relationship for the inductor

$$v(t) = L\frac{di(t)}{dt}$$

has a Laplace-domain counterpart

$$V(s) = LsI(s).$$

According to the Laplace transform table, if the initial condition is not zero, the Laplace-domain counterpart is

$$V(s) = L[sI(s) - i(0)] = LsI(s) - Li(0) = Ls\left[I(s) - \frac{i(0)}{s}\right].$$

There are two ways to handle this initial condition. The first way is to use the relationship

$$V(s) = LsI(s) - Li(0)$$

and treat the initial condition as a voltage source of value

$$-Li(0) = -3.$$

The second way is to use the relationship

$$I(s) = \frac{V(s)}{Ls} + \frac{i(0)}{s}$$

and treat the initial condition as a current source of value

$$\frac{i(0)}{s} = \frac{3}{s}.$$

Therefore, we can have two Laplace-main circuits, as shown in Fig. S23.3a and Fig. S23.3b, respectively.

Fig. S23.3a

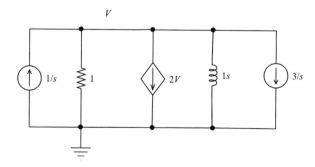

Fig. S23.3b

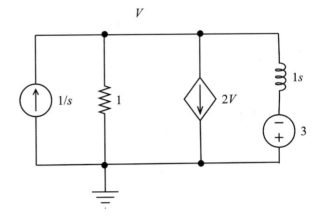

V

Let us first solve the circuit in Fig. S23.3a. The node equation is

$$\frac{V}{1} + \frac{V}{s} = \frac{1}{s} - 2V - \frac{3}{s},$$

$$V = -\frac{2}{3\left(s + \frac{1}{3}\right)}.$$

The current through the inductor is $i(t)$ and its Laplace transform is

$$I = \frac{V}{s} = -\frac{2}{3s\left(s + \frac{1}{3}\right)} = \frac{-2}{s} + \frac{2}{s + \frac{1}{3}}.$$

After taking the inverse Laplace transform using the Table of Laplace Transform Pairs,

$$i(t) = -2 + 2e^{-\frac{1}{3}t} \text{ A, for } t \geq 0.$$

Now let us solve the circuit in Fig. S23.3b. The node equation is

$$\frac{V}{1} + \frac{V + 3}{s} = \frac{1}{s} - 2V,$$

$$V = -\frac{2}{3\left(s + \frac{1}{3}\right)}.$$

This expression is exactly the same as that for Fig. S23.3a. Therefore, we have the same solution.

Chapter 24. Fourier Transform in Circuit Analysis

Problem 24.1 Use the Fourier transform method to solve for the circuit in Fig. P24.1. The initial voltage in the capacitor is 3 V.

Fig. P24.1

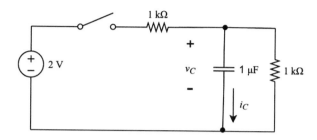

Solution

This problem is the same as Problem 23.2, in which the Laplace transform method is used. The difficult part of solving this problem using the Fourier transform is that the Fourier transform method does not have an explicit way to handle the initial conditions.

In order to accommodating an initial condition, we need to introduce a source with a step function as follows.

In Problem 23.2, the initial condition of the capacitor is treated as a voltage source of value in the Laplace domain

$$\frac{v(0)}{s} = \frac{3}{s}.$$

Converting this source into the time domain gives the time-domain source

$$v(0)u(t) = 3\,u(t),$$

where $u(t)$ is the unit step function. The Fourier transform of the unit step function can be found in the Table of Fourier transform Pairs as

$$\pi\delta(\omega) + \frac{1}{j\omega}.$$

We can redraw the circuit by introducing the step function sources in Fig. S24.1.

We now set up a node equation for the circuit in Fig. S24.1 in the Fourier domain as follows.

Fig. S24.1

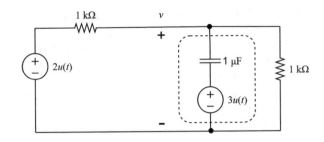

$$\frac{V - 2\left[\pi\delta(\omega) + \frac{1}{j\omega}\right]}{1k} + \frac{V - 3\left[\pi\delta(\omega) + \frac{1}{j\omega}\right]}{\frac{1}{j\omega\mu}} + \frac{V}{1k} = 0,$$

$$V - 2\left[\pi\delta(\omega) + \frac{1}{j\omega}\right] + j\omega m V - 3j\omega m\left[\pi\delta(\omega) + \frac{1}{j\omega}\right] + V = 0,$$

$$V = \frac{2\left[\pi\delta(\omega) + \frac{1}{j\omega}\right] + 3j\omega m\left[\pi\delta(\omega) + \frac{1}{j\omega}\right]}{2 + j\omega m},$$

$$V = \frac{2000\left[\pi\delta(\omega) + \frac{1}{j\omega}\right] + 3j\omega\left[\pi\delta(\omega) + \frac{1}{j\omega}\right]}{2000 + j\omega},$$

$$V = \frac{k_1}{2000 + j\omega} + k_2\left[\pi\delta(\omega) + \frac{1}{j\omega}\right].$$

It can be verified that $k_1 = 2$ and $k_2 = 1$. After taking the inverse Fourier transform by using the Table of Fourier Transform Pairs,

$$v(t) = 1 + 2e^{-2000t} \text{ V, for } t \geq 0.$$

This answer is the same as that obtained from Problem 23.2, obtained by the Laplace transform. The voltage is the capacitor voltage, which is normally denoted by v_C, as in Fig. P24.1. This voltage includes the everything inside the dotted box in Fig. S24.1.

We feel that if a circuit problem involves initial conditions or switch actions, the Laplace transform method is easier than the Fourier transform method because the Laplace transform of the unit step function is $1/s$, while the Fourier transform of the unit step function is $\pi\delta(\omega) + 1/(j\omega)$.

Problem 24.2 The input signal is a signum function. Find the inductor current i_L.

Fig. P24.2

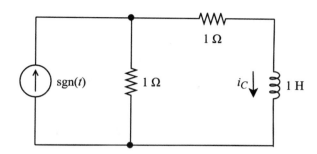

Solution

This problem cannot be solved by the Laplace transform method because the input signal is not zero when $t < 0$.

This problem cannot be solved by the phasor method either because the input signal is not sinusoidal.

This problem can be solved either in the time domain or in the Fourier domain. We will use the Fourier transform method here.

The Fourier transform of the signum function is

$$\frac{1}{j\omega}.$$

We notice that this circuit is current divider. Let us set up the current divider relationship in the Fourier domain as follows.

$$I_C = \left(\frac{1}{j\omega}\right)\frac{1}{(1+j\omega)+1} = \left(\frac{1}{j\omega}\right)\frac{1}{2+j\omega} = \frac{1/2}{j\omega} - \frac{1/2}{2+j\omega}.$$

Taking the invers Fourier transform yields

$$i_C(t) = \frac{1}{2}\operatorname{sgn}(t) - \frac{1}{2}e^{-2t}u(t) \text{ A},$$

where $u(t)$ is the unit step function.

Problem 24.3 Repeat Problem 24.2 with $10\cos(4t)$ being the input signal.

Solution

Since the input function is sinusoidal, both the phasor method and the Fourier method can be used. The Laplace transform method does not work.

The Fourier transform of the input function is

$$10\pi[\delta(\omega + 4) + \delta(\omega - 4)].$$

Let us set up the current divider relationship in the Fourier domain as follows.

$$I_C = (10\pi[\delta(\omega + 4) + \delta(\omega - 4)])\frac{1}{(1 + j\omega) + 1},$$

$$= 10\pi\left[\frac{\delta(\omega + 4)}{2 + j\omega} + \frac{\delta(\omega - 4)}{2 + j\omega}\right],$$

$$= 10\pi\left[\frac{\delta(\omega + 4)}{2 + j(-4)} + \frac{\delta(\omega - 4)}{2 + j(4)}\right],$$

$$= \pi[(1 + j2)\delta(\omega + 4) + (1 - j2)\delta(\omega - 4)],$$

$$= \pi[(\delta(\omega + 4) + \delta(\omega - 4)] + 2j\pi[(\delta(\omega + 4) - \delta(\omega - 4)].$$

Taking the invers Fourier transform yields

$$i_C(t) = \cos(4t) - 2\sin(4t) \text{ A.}$$

Chapter 25. Second-Order Circuits

Problem 25.1 In a series RLC circuit, what does the step response look like when $R = 0$? The input is step voltage source, and the output is the voltage across the capacitor.

Fig. P25.1

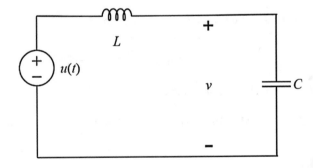

Solution
For a step response, it is easier to the Laplace transform method. The Laplace-domain circuit is shown in Fig. S25.1.

Fig. S25.1

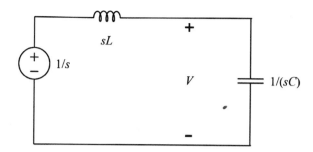

This circuit is a voltage divider. We have

$$V = \left(\frac{1}{s}\right) \frac{\frac{1}{sC}}{sL + \frac{1}{sC}} = \left(\frac{1}{s}\right) \frac{\frac{1}{LC}}{s^2 + \frac{1}{LC}} = \frac{1}{s} - \frac{s}{s^2 + \frac{1}{LC}}.$$

Taking the inverse Laplace transform yields

$$v(t) = \left[1 - \cos\left(\frac{1}{\sqrt{LC}} t\right) \right] u(t).$$

This function is biased sinewave after $t = 0$. When R is 0, the series RLC circuit becomes an oscillator. The effect of the resistance R is to damp the oscillation. A larger R has a stronger damping effect.

Problem 25.2 In a parallel RLC circuit, what does the step response look like when $R = \infty$? The input is step current source, and the output is the current through the inductor.

Fig. P25.2

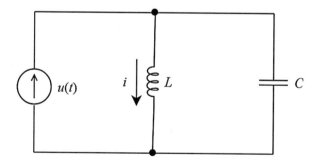

Solution
For a step response, it is easier to the Laplace transform method. The Laplace-domain circuit is shown in Fig. S25.2.

Fig. S25.2

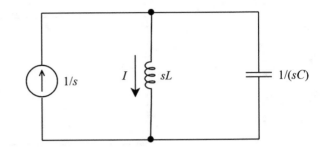

This circuit is a current divider. We have

$$I = \left(\frac{1}{s}\right)\frac{\frac{1}{sC}}{sL + \frac{1}{sC}} = \left(\frac{1}{s}\right)\frac{\frac{1}{LC}}{s^2 + \frac{1}{LC}} = \frac{1}{s} - \frac{s}{s^2 + \frac{1}{LC}}.$$

Taking the inverse Laplace transform yields

$$i(t) = \left[1 - \cos\left(\frac{1}{\sqrt{LC}}t\right)\right]u(t).$$

This function is biased sinewave after $t = 0$. When R is ∞, the parallel RLC circuit becomes an oscillator. A smaller LC gives a higher oscillating frequency. The effect of the resistance R is to damp the oscillation. A smaller R has a stronger damping effect.

Problem 25.3 The circuit in Fig. P25.3 is neither a series nor a parallel circuit. It is still a second-order RLC circuit. Find the conditions for circuit to be underdamped, critically damped, and overdamped.

Fig. P25.3

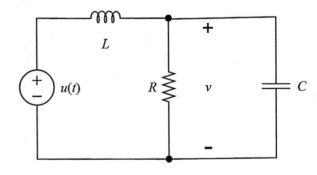

Solution

For a step response, it is easier to the Laplace transform method. The Laplace-domain circuit is shown in Fig. S25.3.

Fig. S25.3

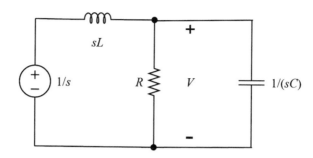

This circuit is a voltage divider. We have

$$V = \left(\frac{1}{s}\right)\frac{\frac{\frac{R}{sC}}{R+\frac{1}{sC}}}{sL + \left(\frac{\frac{R}{sC}}{R+\frac{1}{sC}}\right)} = \left(\frac{1}{s}\right)\frac{\frac{R}{sC}}{sL\left(R+\frac{1}{sC}\right) + \left(\frac{R}{sC}\right)} = \left(\frac{1}{s}\right)\frac{R}{s^2RLC + sL + R}.$$

In the above equation, the input is $1/s$ and the transfer function is

$$H = \frac{R}{s^2RLC + sL + R} = \frac{\frac{1}{LC}}{s^2 + s\frac{1}{RC} + \frac{1}{LC}}.$$

Letting the denominator of the transfer function be 0, we have the characteristic equation of the circuit

$$s^2 + s\frac{1}{RC} + \frac{1}{LC} = 0.$$

The solutions of this characteristic equation are given as

$$s_{1,2} = \frac{-\frac{1}{RC} \pm \sqrt{\frac{1}{(RC)^2} - \frac{4}{LC}}}{2}.$$

The condition for a **critically damped** circuit is when the two solutions are identical, that is,

$$\frac{1}{(RC)^2} - \frac{4}{LC} = 0,$$

$$R^2 = \frac{L}{4C}.$$

The condition for an **overdamped** circuit is when the two solutions are two different negative real numbers. First, we require

$$\frac{1}{(RC)^2} - \frac{4}{LC} > 0,$$

$$R^2 < \frac{L}{4C}.$$

The condition for an **underdamped** circuit is when the two solutions are a pair of complex conjugate numbers.

$$\frac{1}{(RC)^2} - \frac{4}{LC} < 0,$$

$$R^2 > \frac{L}{4C}.$$

When the circuit is overdamped or critically damped, the stability requirement is that the solutions be negative, that is,

$$-\frac{1}{RC} + \sqrt{\frac{1}{(RC)^2} - \frac{4}{LC}} < 0,$$

$$\sqrt{\frac{1}{(RC)^2} - \frac{4}{LC}} < \frac{1}{RC},$$

$$\frac{1}{(RC)^2} - \frac{4}{LC} < \frac{1}{(RC)^2},$$

$$0 < \frac{4}{LC},$$

which is always valid.

Our conditions are the same as those for the parallel RLC circuits.

Chapter 26. Filters

Problem 26.1 Without doing any mathematical derivation, determine whether the Sallen-Key filter shown in Fig. P26.1 a lowpass filter or a highpass filter.

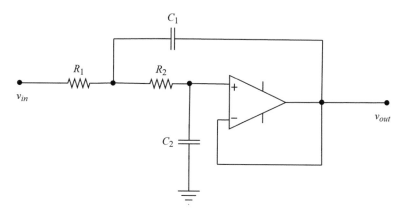

Fig. P26.1

Solution
Let us consider two extreme cases.

When the frequency is very low, the capacitors are open circuits. Figure P26.1 becomes Fig. S26.1a.

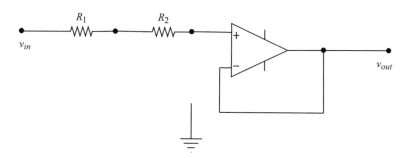

Fig. S26.1a

This circuit is a voltage follower (also known as a buffer) with

$$v_{out} = v_{in}.$$

When the frequency is very high, the capacitors are short circuits. Figure P26.1 becomes Fig. S26.1b.

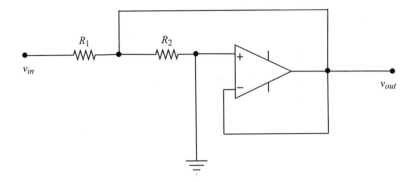

Fig. S26.1b

In Fig. S26.1b, the non-inverting input of the op-amp is shorted to the ground. As a result, the inverting input and the output also have the voltage of 0.

Therefore, this is a lowpass filter.

Problem 26.2 Without doing any mathematical derivation, determine whether the Sallen-Key filter shown in Fig. P26.2 a lowpass filter or a highpass filter.

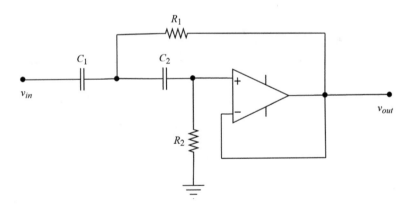

Fig. P26.2

Solution

Let us consider two extreme cases.

When the frequency is very low, the capacitors are open circuits. Figure P26.2 becomes Fig. S26.2a.

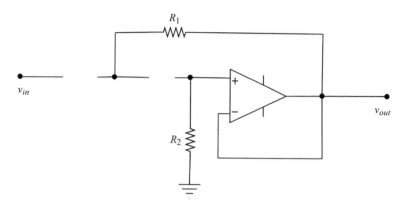

Fig. S26.2a

In this circuit, the input is not fed into the circuit. The op-amp's inverting and non-inverting inputs are 0. Therefore, the output is 0.

When the frequency is very high, the capacitors are short circuits. Figure P26.2 becomes Fig. S26.2b.

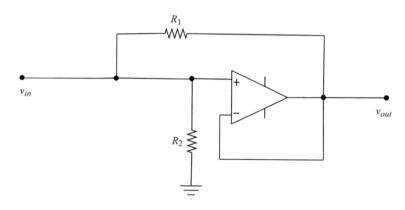

Fig. S26.2b

Figure S26.2b is a voltage follower,

$$v_{\text{out}} = v_{\text{in}}.$$

Therefore, this is a highpass filter.

Problem 26.3 Without doing any mathematical derivation, determine whether the Sallen-Key filter shown in Fig.P26.2 a lowpass filter or a highpass filter or none of them.

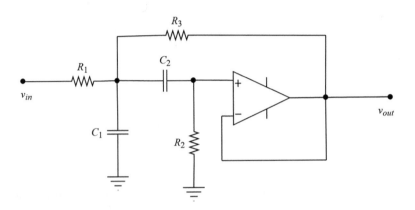

Fig. P26.3

Solution
At very low frequency, the capacitors are open circuits. Figure P26.3 becomes Fig. S26.3a.

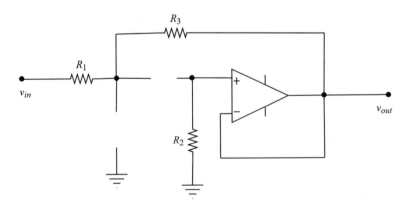

Fig. S26.3a

In this circuit, the op-amp's inverting and non-inverting inputs are 0. Therefore, the output is 0.

At very high frequency, the capacitors are short circuits. Figure S26.3a becomes Fig. S26.3b.

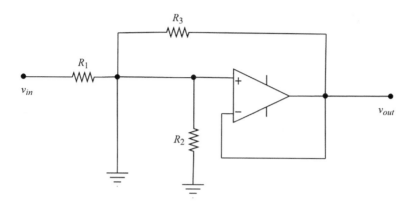

Fig. S26.3b

Once again, in Fig. S26.3b, the op-amp's inverting and non-inverting inputs are 0. Therefore, the output is 0.

This filter has 0 output at very low frequency and very high frequency. It is not a lowpass filter and is not a highpass filter, either. We observe that there are two capacitors in the circuit. This circuit is most likely a second-order bandpass filter.

Chapter 27. Wrapping Up

Problem 27.1 Are KVL equations and mesh equations the same? Are KCL equations and node equations the same.

Solution

Mesh equations are KVL equations and are the special applications of the KVL principle. A circuit may have many elements. If we use the element's currents and voltages as variables, we need a lot of equations to solve them. In principle, if we have 20 variables, we need 20 equations.

A circuit may only have few meshes. Thus, we only need to solve a system of a small number of mesh equations. Once the mesh currents are obtained, the element's currents and voltages can be readily evaluated using the mesh currents.

Similarly, node equations are the special applications of the KCL principle. A circuit may have few essential nodes. An essential node is a node joining three or more elements. Thus, we only need to solve a system of a small number of node equations. Once the node voltages are obtained, the element's currents and voltages can be readily evaluated using the node voltages.

The biggest motivation to use node equations or mesh equations is to reduce the number of equations to its minimum.

Problem 27.2 Do you have a preference regarding to nodes equations or mesh equations?

Solution
In principle, if the number of essentials nodes is greater than the number of meshes, use the mesh equation; otherwise, use the node equations. You commonly have enough equations to solve for the circuit. You do not need to use both mesh equations and node equations to start with circuit analysis.

There are many other strategies. One strategy is only set up equations for the inductor currents and capacitor voltages. Once the inductor currents and capacitor voltages are obtained, other currents and voltages can be easily solved.

Problem 27.3 The main purpose of the Laplace transform and the Fourier transform is to avoid solving differential equations in the time domain. Why do we need both the Laplace transform and the Fourier transform? Can we just learn one of them?

Solution
The applications of the Laplace transform and Fourier transform have many overlaps. In many cases, you can use either the Laplace transform or the Fourier transform. By using Laplace transform and Fourier transform, you can treat capacitors and inductors in the same way as you treat resistors.

The Laplace transform is easier to work with if the circuit involves initial conditions and switch actions. The Laplace transform is easier to use for time transient analysis.

The Laplace transform assumes that the signals are zero when $t < 0$. On the other hand, the Fourier transform does not have this restriction. The Fourier transform assumes zero initial conditions. If we have to use the Fourier transform to deal with a circuit with initial conditions, we need to artificially introduce some step sources into the capacitors and inductors.

Problem 27.4 Which is more powerful, the Fourier transform method or the phasor method?

Solution

The Fourier transform method is more powerful and has wider applications. The phasor method can only handle one frequency. The Fourier transform method can solve all the problems that the phasor method can solve. On the other hand, the phasor method is easier to use in applications where the frequency is always fixed, for example, in the power system.

Problem 27.5 Name one most important concept in electric circuits.

Solution

In our opinion, it is Ohm's law.

Bibliography (Some Textbooks Used in Colleges)

Basic Engineering Circuit Analysis, 11th Ed., J. David Irwin and R. Mark Nelms, John Wiley and Sons. ISBN: 978-1-118-53929-3, 2015.

Circuit Analysis and Design, Fawwaz T. Ulaby, Michel M. Maharbiz and Cynthia M. Furse, Michigan Publishing, 2018

Electric Circuits, 11th Edition, James W. Nilsson and Susan Reidel, Pearson, ISBN-139780134746968, 2019

Engineering Circuit Analysis, 9th Edition, William Hayt, Jack Kemmerly, and Steven Durbin, McGraw Hill, ISBN13: 9780073545516, 2019.

Essentials of Electrical and Computer Engineering, David V. Kerns, Jr. and J. David Irwin, Pearson Prentice Hall, Upper saddle River, ISBN-13: 978-0139239700, 2004

Fundamentals of Electric Circuits, 6th edition, Charles K. Alexander and Matthew N. O. Sadiku, Mc Graw Hill, ISBN-13: 978-0078028229, 2017

Fundamentals of Electronic Circuit design, David J. Comer and Donald T. Comer, Wiley, ISBN-13: 978-0471410164, 2002

Introduction to Electric Circuits, James A. Svoboda and Richard C. Dorf, John Wiley & Sons Inc., NY, 9th Edition, 2013. ISBN 1118477502,

The Analysis and Design of Linear Circuits, 7th edition, Roland E. Thomas, Albert J. Rosa and Gregory J. Toussaint, John Wiley & Sons, Inc.

The Analysis and Design of Linear Circuits (electronic version OK), Roland E. Thomas and Albert J. Rosa and Gregory J. Toussaint, Wiley, ISBN 978-1-118-06558-7, 2012.

© The Editor(s) (if applicable) and The Author(s), under exclusive license to Springer Nature Switzerland AG 2021
G. L. Zeng, M. Zeng, *Electric Circuits*, https://doi.org/10.1007/978-3-030-60515-5

Index

© The Editor(s) (if applicable) and The Author(s), under exclusive license to
Springer Nature Switzerland AG 2021
G. L. Zeng, M. Zeng, *Electric Circuits*, https://doi.org/10.1007/978-3-030-60515-5

Printed in the United States
by Baker & Taylor Publisher Services